SpringerWienNewYork

Studies in Space Policy
Volume 4

Edited by the European Space Policy Institute
Director: Kai-Uwe Schrogl

Wolfgang Rathgeber, Kai-Uwe Schrogl,
Ray A. Williamson (eds.)

The Fair and Responsible Use of Space: An International Perspective

SpringerWienNewYork

Wolfgang Rathgeber
Kai-Uwe Schrogl
Ray A. Williamson

Softcover reprint of the hardcover 1st edition 2010

SpringerWienNewYork is a part of
Springer Science + Business Media
springer.at

Typesetting: Thomson Press (India) Ltd., Chennai
Printing: Strauss GmbH, 69509 Mörlenbach, Germany

Cover: ESA, "Schematic depiction of Earth Observation satellites orbits"

Printed on acid-free and chlorine-free bleached paper

SPIN: 12745265

With 35 Figures

Library of Congress Control Number: 2010926411

ISBN 978-3-211-99652-2 SpringerWien NewYork

Preface

The International Academy of Astronautics (IAA) pursues the goal of developing astronautics for peaceful purposes. Approaching its 50th year of existence, it recognises individuals who have distinguished themselves in a related branch of science and technology through full or corresponding membership. Currently, there are more than 1200 members from 75 countries, representing outstanding and renowned experts in their respective field of work. Academy members contribute to international endeavours and cooperate in the advancement of aerospace science.

Academy activities are supported by various scientific symposia and meetings as well as by the work of six specialised commissions. A special role is played by the "Cosmic Studies" and "Position Papers" on relevant issues in various areas of the Academy's scope. Further important publications are the Acta Astronautica, Proceedings of Symposia, Dictionaries, a Yearbook and the IAA e-newsletter. These communications cover a broad range of topics and they reflect the Academy's intention to serve as a discussion platform for different questions within the Academy's domain of interest.

One issue area that is gaining in importance is the fair and responsible use of outer space, given that space applications are at the heart of everyday interaction and communication. More and more actors are becoming involved in space activities, often with a lack of coordination and coherence. There is a need to investigate how an equitable, just and sustainable utilisation of space can be achieved for both established and new space actors. In this regard, aspects related to space benefits for developing countries, the peaceful use of outer space and Space Situational Awareness need to be considered. Moreover, the role of the United Nations as well as questions of data sharing and Space Traffic Management have to be discussed. This has to be complemented by deliberations on governance issues of the international system as well as by new thoughts on how to shape international cooperation.

Against this background, the IAA has patronaged and co-sponsored the conference on "The Fair and Responsible Use of Space: An International Perspective". The event took place in November 2008, and it was organised jointly with the European Space Policy Institute (ESPI), which was also the local host in Vienna, and the Secure World Foundation (SWF). These institutions brought in their resources, their expertise and their experience, creating the added value and synergy needed to adequately approach such a multi-layered topic. For the Academy, it was a pleasure to see that numerous of its members acted as speakers or moderators.

Preceding this book, ESPI published the document "10 Steps to Achieve Fair and Responsible Use of Space" as another immanent outcome of the conference. This text will shortly be translated into an official "Position Paper" of the IAA. The Academy considers the whole project a highly successful one and sees it as an excellent basis for future efforts in this critical domain.

Jean-Michel Contant
Secretary General
International Academy of Astronautics

Foreword by the Editors

The issue of a sustainable conduct of space activities is gaining importance, since modern societies depend on space and its applications. Everyday life would be seriously degraded, if not impossible, without the utilisation of space-based science and technology. This holds true for the present generation, but also for the ones to come. Accordingly, space has to be preserved for the future. Sustainability can be achieved through a fair and responsible use of space. Given the growing number of space actors and their diverse nature, the time to think and act is now.

This will involve a number of interrelated issues. First of all, relevant facts and background information like the current state of the space environment and its potential degradation will have to be considered. Moreover, concrete rules facilitating the fair and responsible use of space will need to be devised and enforced. Finally, ways to involve all stakeholders and to achieve engagement at a global scale will have to be found. These issue areas cannot be tackled separately from each other, but will have to be treated in a unified and cross-cutting way.

With these thoughts in mind, the European Space Policy Institute (ESPI) hosted a conference on "The Fair and Responsible Use of Space: An International Perspective" on 20–21 November 2008 at Vienna. The conference was organised jointly with the International Academy of Astronautics (IAA) and the Secure World Foundation (SWF). Selected international experts from various backgrounds gave presentations and engaged in discussions with the audience of more than 60 high-ranking professionals.

At the conference, the associated questions were addressed in three different sessions – the first one dealing with pertinent policy issues, the second one with suitable space traffic rules and the third one with the way to a truly global engagement. In this framework conclusions and policy recommendations on aspects related to space benefits for developing countries, the peaceful use of outer space and Space Situational Awareness were presented. The role of the UN as well as data sharing issues and Space Traffic Management were also addressed. In addition, a suitable structure for the international system was debated.

As one of the results, the conference came up with proposals for action in the form of the "10 Steps to Achieve Fair and Responsible Use of Outer Space". Furthermore, the discussions at the conference served as a starting point for devising articles based on the contributions by the presenters and the feedback received from the audience. This book gathers these value-added articles. In addition, it features extra material such as complementary considerations and relevant information. Through this content, the book will stimulate and sustain the debate on the crucial of question how humankind could and should use outer space in a fair and responsible manner – serving the needs of the present generation without impairing future generations' ability to conduct space activities.

Wolfgang Rathgeber, Kai-Uwe Schrogl, Ray A. Williamson

Table of contents

1 The general concepts of fairness and responsibility

Wolfgang Rathgeber

This book is concerned with the fair and responsible use of outer space. To provide a conceptual basis for the discussion of various szpace-related issues, the two concepts of fairness and responsibility will be discussed, first in a general way and then they will be treated separately, including consideration of areas where they overlap. By setting out their contents, possible interlinkages and consequences for different domains, a conceptual framework is provided for the topics that will be treated subsequently. After this theoretical approach, the following chapter will apply the concepts of fairness and responsibility specifically to space activities.

1.1. Definition of terms

The nature and application of the concept of fairness have been critical questions throughout human history. Despite the considerable philosophical and other attention paid to the concept, it remains difficult to identify a single definition that is universally understood, accepted and applicable to all domains where fairness is an issue. Generally speaking, however, fairness refers to norms that facilitate a reasonable coexistence within societies. As such, fairness is an important prerequisite of peace.

Fairness is closely associated with justice and both terms are sometimes used synonymously. The latter, though, has a more legal connotation – it is often understood as the impartiality of the law. In art, this impartiality of the law is often symbolised by a blindfolded goddess of justice with scales in her hands, as shown in Figure 1. In philosophical discourse, the term justice appears more often than fairness. Both fairness and justice are concerned with assigning goods or opportunities on the basis that everybody gets his or her due share.

The concept of fairness and justice can be applied to both processes and to distributions. Regarding processes, fairness and justice refer to institutions, mechanisms and policies. As for distributions, fairness and justice are concerned with the amounts, quotas and ratios to be allocated. A theory of the latter, often described as distributive justice, was presented by the 20th century philosopher

Fig. 1. *The impartiality of law (source: sculpturegallery).*

John Rawls. Some of the basic thoughts are stated in the following because matters of distribution are of special relevance for the topics to be covered in this book. The considerations are taken from Rawls' book "A Theory of Justice".[1]

1.2. Fairness, justice and John Rawls

According to Rawls, justice is the primary virtue of social institutions, just like truth is the basic scale for any system of thought. Justice provides invulnerability to every human being. This invulnerability cannot be set apart, even if it was for the sake of the whole society's well-being. It is not acceptable to demand sacrifices from a few in order to bring about a better situation for the vast majority. In this regard, Rawls' approach is in contrast to strict utilitarianism. Rawls also maintains that equal rights are necessary in a just society. The only reason to accept injustice is to avoid a situation where more severe injustice would result.

To establish a rational basis for these intuitive deliberations about the priority of justice, Rawls sets forth a theory of justice starting from a model of societies. According to him, societies are unions of humans that accept some rules as binding for their relations. These rules support a system of cooperation aimed at the well-being of its participants. Such societies are meant to promote mutual advantage, but they are not free from conflict because everybody strives to have more of the goods to be distributed.

As a consequence, principles are needed to choose between the various possible ways of allocating goods. These principles, which enable the assignment of rights and duties within the most important institutions of the society and decisions about the right distribution of gains and losses of societal cooperation, are the basis of social justice. Rawls calls societies well-ordered if in addition to being aimed at the well-being of their members, they are also effectively steered by a common perception of justice. This common perception of justice creates peace among society members with different aims because the general striving for justice limits the pursuit of other goals.

Real-world societies, however, are rarely well-ordered because there is a continual debate on the nature and content of justice and injustice. Still, one can assume that while all members of a society may each have their individual ideas about justice, they each see the necessity of basic principles for social justice as described above. Consequently, humans with different ideas about justice can still agree that institutions are fair if basic rights and duties are not assigned on an arbitrary basis and if the rules establish a sensible trade-off between conflicting interests for the sake of society's welfare.

For a human society to work, though, the prevalent ideas of justice have to allow for a certain overlap with other issues that need to be considered. These include coordination, efficiency and stability. Regarding coordination, the different tasks, plans and endeavours of humans have to be accounted for in a way that allows their simultaneous execution without significantly impairing each other. As for efficiency, the execution of plans has to achieve societal goals in a way that occupies or consumes as few resources as possible. Finally, stability refers to a situation where the scheme of societal cooperation is followed on a continuous basis and where deviations are countered. These issues again are connected to the principle of justice.

Seen as a whole, justice is not just concerned with rights, duties and distributions. In supporting the functioning of societies, it also has to account for related issues and secondary effects which in turn feed back upon the underlying notion of justice. Existing ideas of justice differ not only in the distribution they foresee, but also in the wide-scale consequences they bring about. In fact, it is the latter aspect that can provide scales for evaluating various approaches to justice as long as the wide-scale consequences are quantifiable.

Following these considerations about the role of justice, Rawls lays down a main concept, namely that of *justice as fairness*. To this end, he generalises the theory of the social contract of Locke, Rousseau and Kant, claiming that the social contract mainly refers to the principles of justice on which the basic

Fig. 2. *John Rawls (source: Harvard University Gazette).*

structure of society is based. These principles are seen to be those that free and reasonable humans would agree upon to govern their basic relations for the sake of their own good in an initial situation of equality, the so-called original position. This original position, of course, is not conceived as a real situation but rather as a theoretical construct (Figure 2).

One of the features of the original position is a lack of knowledge about the situation, class, status and physical properties of each member of the society. The basic principles of justice are set behind a "veil of ignorance". This ensures that the accidental character of nature or societal status does not cause advantages or disadvantages to society's members. If nobody knows his own specific features, nobody will be able to think about principles that put him in a favourable position with regard to others. Thus, the resulting norms could be considered fair, justifying the notion of *justice as fairness*. However, Rawls stresses the fact that justice does not equal fairness.

The core problem is to find the principles that a society in an original position would agree upon. In this context Rawls takes a stand against the principle of utility, claiming it is not likely that equal members of a society opt for norms that decrease the prospects of living for some in order to maximise the welfare of others. In his view, the principle of utility is not compatible with the cooperation of equals aiming at mutual benefit, which is characteristic of a well-ordered society.

Rather, humans in the original position would opt for two basic principles: (i) one consisting of equal rights and duties and (ii) the other holding that societal or economic inequalities are just and acceptable only if they lead to advantages for everybody, in particular for the weakest members of the society. This is different to the situation earlier mentioned where greater total welfare justifies the unfavourable situation of some. The two basic principles follow from a perception of justice where accidental fluctuations in talent or societal circumstances do not lead to political or economic advantages. The theory of justice as fairness draws upon two elements that are separable from each other: (i) one dealing with the description of the original position, including the decision problem and (ii) the second dealing with the set of principles that supposedly would be agreed upon.

Rawls also dwells upon distributions as such, emphasising that political and economic systems do not only account for needs that exist independently. Rather, these systems also generate future needs by themselves. Consequently, the allocation of goods cannot be seen independently from the institutions of the society in which goods are to be distributed. Institutions are understood to be

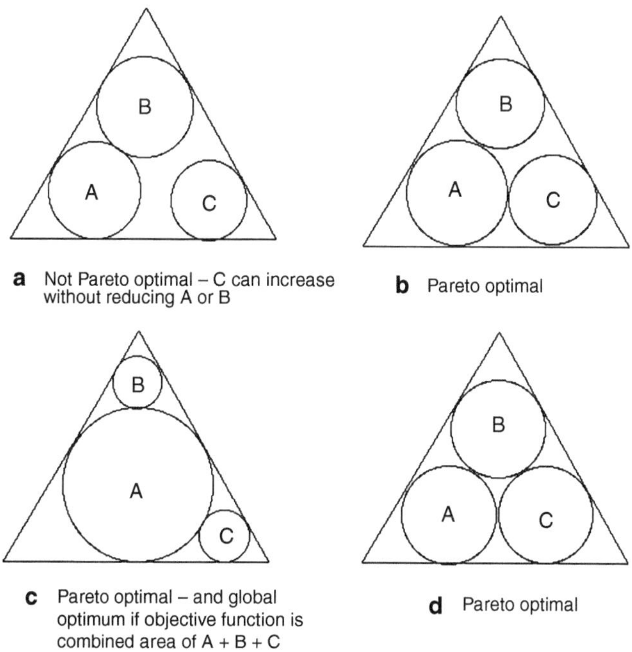

Fig. 3. *Pareto optimum (source: Standford University).*

public systems of regulation that define the rights and duties, powers and immunities of positions and offices.

An important criterion in allocating goods is the Pareto optimum. It deals with a given set of goods to be distributed. A specific assignment of goods is called optimal in a Pareto sense if there is no other distribution of goods that puts at least one member of the society into a more favourable position while leaving the position of the other members unchanged. Figure 3 illustrates the concept of the Pareto optimum. Rawls claims that this is the scale by which a society in its original position would judge the efficiency of the respective mechanisms. This concludes the description of some ideas laid down in “A Theory of Justice”.

A final point to be made concerns distributions. In allocating goods or opportunities, the basis to be chosen is a crucial point. One potential basis is constituted by equality, meaning that everybody gets the same share. Another potential basis is given by the needs, both of the donor and the receiver. This requires an unbiased way to measure the need. A third potential basis is given by contribution, supposing that the more somebody feeds into the system, the more he is entitled to receive in return. Again, this raises the question of quantifying the contribution – if this is possible in the first place. Contributions can be measured in an absolute way, allowing for direct comparison to other contributions. They can also be measured against the available capacity of the contributor, giving a more appropriate picture of the efforts he takes and the relative value he provides.

1.3. The meaning of responsibility

Having looked at the principles of justice and fairness in some detail, some consideration will now be paid to the concept of responsibility, which is the second key word in this book’s title. Responsibility may be defined as using power to act in a way that respects existing rights, duties and needs and that allows for long-term benefits. Power in this regard comprises favourable conditions such as physical strength, economic wealth, technical capacity, intellectual ability or empirical knowledge.

Responsibility is not restricted to individuals but is also applicable to groups of individuals or societies. Responsibility can be realised towards oneself, in which case one’s own duties are respected (one’s own rights are not relevant in this regard). It is then the consequence of a rational and reasonable attitude. Responsibility can also be realised towards others, respecting their possibly different rights and needs. Responsibility towards others is a consequence of a moral attitude. It is obvious that responsibility towards oneself and responsibility towards others are not clearly separable.

The benefits resulting from responsible behaviour can accrue to oneself, or to others or, ideally, to both. In this regard, responsibility is a precondition for sustainability. Benefits are understood as the sum of advantages outweighing the sum of disadvantages. However, the disadvantages have to be distributed in a fair manner. This prerequisite is enshrined in the concept of respect for needs and rights. Otherwise, the approach would be utilitarian: the sacrifices of some would be acceptable as long as the welfare of society as a whole is maximised.

Responsibility again is a concept consisting of two elements that can be separated from each other. The first one comprises analysing and morally valuing the likely consequences of each available path of action over time. The second one is the deliberate and conscious decision to choose the path of action associated with the most favourable consequences. Responsible action can only result from the successful implementation of both elements in a combined approach. Of course, some ethical credit will still be given if the first step goes wrong, i.e. if a decision is taken based on an erroneous assessment of its consequences.

1.4. Similarities and differences

Having introduced the notions of justice, fairness and responsibility, common features and interlinkages between them will now be addressed. Both the aims and the results of acting fairly and responsibly will be considered. A further step then is to develop possible criteria or at least classes thereof that allow for operationalisation of the two concepts, with a view to taking them out of the academic context and applying them to situations occurring within the real-world conditions of the space community.

Fairness and responsibility show common features in various regards. First of all, they relate to both individuals and society. This means that they are applicable to both individuals and societies and that they give a basis to the relationship between the two. This basis can be considered essential for the functioning of an association of humans. Another common feature is the inherent necessity to think about others. This is not due to altruistic motives – at least not primarily. Rather, it is due to the fact that rights, needs and attitudes of others need to be incorporated into any model of a society if this model is to lead to satisfactory results. Justice and fairness prevent excessive selfishness, although neither of them forbids a certain amount of egoism per se; in fact, the concept of justice as fairness relies on a healthy amount of egoism among the members of a society.

There are also aspects of fairness and responsibility that can be read as both similar and different. Both concepts aim at avoiding negative consequences of inequality. However, they use different mechanisms to reach that aim. Fairness strives to avoid unequal starting conditions that lead to unfavourable results for some in the allocation of goods by emulating a state of assumed equality. Responsibility, on the other hand, takes existing inequalities as a given fact and tries to account for them. It is important to stress that neither fairness nor responsibility neglects the differences; they just view and handle them in their own specific way.

Responsibility and justice as fairness also contribute to each other. Justice as fairness supports responsible behaviour by providing a basis for stability that allows for assessing different paths of action and pertinent consequences. Responsible behaviour facilitates justice as fairness, because members of the society will be more apt to accept rules that limit their freedom if they know that all other members think about the possible consequences their actions might cause and restrict themselves accordingly. The two concepts of responsibility and justice as fairness are closely interwoven; in fact, it is hard to imagine one without the other.

1.5. Summary and application to the space arena

It is challenging to transfer the ideas of justice, fairness and responsibility to the present situation of the global space community. This is particularly true for the concept of justice as fairness, since the current situation is marked by a strongly differentiated space society with considerably unequal goals, capacities and needs. Accordingly, the theoretical and ideal situation of a society in an original position is hard to imagine. While the kinds of "persons" that would need to be part of the initial state can be identified with some surety: national and international organisations, public and private entities, commercial and non-profit actors; it would have to be assumed that their strengths, weaknesses and specific interests such as prestige, development, knowledge, security or profit are hidden behind the aforementioned veil of ignorance.

One could also be rather sure about the goods to be allocated. These could be classified into enabling general values and resulting specific gains. The enabling general values would, among others, consist of the basic opportunity to run space activities, unhindered access to space, freedom of action in space, as well as the ability to operate space assets safely and reliably. The resulting specific gains would comprise recognition and reputation in the worldwide community, facilitation of everyday life on Earth, economic progress, scientific advance, enhanced protection of citizens and increased welfare of mankind. These goods correspond to the existing interests in the original position (otherwise they would not be goods), but it is not clear who prioritises which good and who is in the best position for the pursuit of his specific goal. This guarantees an unbiased derivation of just principles.

As was stated in the introductory sections, a basic question is exactly what principles space actors would agree upon if they were in an original position and what consequences would follow for the conduct of space activities. These principles would have to refer to both activities and resources. Fair principles would especially be concerned with activities that impact on others and with scarce resources.

Developed and powerful space actors will not be inclined to agree to principles that diminish the favourable position that they hold. Still, these are the kind of principles looked for. Besides moral considerations, developed and powerful space actors might be brought to agree to such principles based on the conviction that space is inherently global and that a cooperation scheme that is open to any potential space actor facilitates space activities for everybody in the long run.

A first and very basic principle would probably be a general principle, like a fundamental theorem of space activities: each space actor (existing as well as upcoming) shall have the possibility of engaging in space activities for the pursuit of his goals as long as neither the activities nor the goals endanger the possibilities to do so for other space actors now or in the future. This principle obviously has a direct link to issues of sustainability. It would need to be translated into more concrete rules – in fact, this is the current process taking place in relevant fora such as the United Nations Committee on the Peaceful Uses of Outer Space (UNCOPUOS). In the course of this process, various results have been achieved – legally binding instruments such as the Outer Space Treaty (OST) or soft law regulations such as the mitigation guidelines regarding space debris.

Space resources are to be distributed following the allocation of the general goods described above. Such resources can be natural such as adequate launch sites, slots on geosynchronous orbits, radio frequency spectrum bands, or windows of opportunity set by environmental and astronomical conditions. Another class of resources is technical hardware such as spaceports, launchers, space assets and related ground facilities. Perhaps the most important resources are intellectual resources such as engineering know-how, enabling knowledge or supporting information. The data appearing in this context constitutes a special commodity because it can be copied. This means that, in contrast to situations dealing with other goods, a party passing on data does not lose it. The only reasons to not give away data could be the desire to maintain or achieve a leading position and security concerns. This issue will be taken up later in this book.

With this abstract conceptual frame laid down, some topics relevant for the fair and responsible use of outer space will be treated in this book. Before that, though, the next chapter will introduce more concrete definitions for the terms of fairness and responsibility, accounting for their real-world environment of the space community. It should be kept in mind that any operationalisation of the concepts of fairness and responsibility needs to keep in mind the basic considerations stated above.

The topics addressed later in the book are divided into three thematic main blocks, mirroring the structure of the respective IAA/ESPI/SWF conference in November 2008. The first cluster deals with current questions regarding the fair and responsible use of outer space, the second block covers fair rules on orbit and the third block is dedicated to the question of how to achieve global engagement.

[1] Rawls, John. A Theory of Justice. Cambridge: Harvard University Press, 2005.

2 Fairness and responsibility in space activities

Ray A. Williamson

2.1. Introduction

This conference raised many questions about the fair and responsible use of outer space, but two questions seem to me of particular importance: (1) Do all space actors have common understanding of what is meant by "fair and responsible"? and (2) How does the international community ensure that space actors use the precious resource represented by outer space, and especially Earth orbits, responsibly and fairly? In attempting to answer these questions in this conference, I fear that we opened more lines of inquiry than we closed. Nevertheless, the conference presentations and the ensuing discussion, both during the formal proceedings and during the breaks and meals, demonstrated significant progress in sharpening the dialogue towards finding workable solutions to the challenge of the fair and responsible use of outer space.

The English term "fair" conveys a sense of justice, that there is a body of law or common practice to which one can refer in determining exactly how fair an act or collection of acts might be. As applied to space activities, the term "fair" implies that the developed States with major technological resources to call on in accessing and using outer space act in ways that do not impede the ability of less fortunate States to take part in the many benefits that operating in outer space can provide. It might also include the view that the developed States are open to innovations by the developing States in approaching space activities. For some actors, it may further imply that the developed States should assist the less developed ones in accessing the benefits of space activities. This is a basic tenet of the UN Office of Outer Space Affairs (UNOOSA), which works to broaden the capacity of developing States in space technology.

The term "responsible" carries the concept of fair or just behaviour further and implies the possibility of a judgement by the international community as to whether or not the space actor is living up to its duty to maintain certain standards of behaviour, usually defined by established international informal or formal codes. In this case, responsible behaviour implies that developed and developing States must all adhere to the obligation to treat the commons of outer space in such a way as to ensure sustainability of outer space activities into the future.

Within those two simplified answers, of course, lie many qualifications and nuances of meaning.

Until recently, the realm beyond the atmosphere has been treated largely as open territory, a commons belonging to no State, and therefore open to exploitation by any State as it suited. In fact, the metaphor of the American Wild West has often been used in describing the opportunity presented by the use of outer space. In the past few years, the metaphor has been extended to another aspect of outer space: the relative lack of regulation. It is a global commons with little governance beyond the four UN treaties and the regulations of the International Telecommunication Union. States have tended to say, "do what you want in outer space; just stay out of my way". Yet government officials and the general public have come to realise that space activities must be subject to greater caution and care than seemed necessary just a few years ago. This realisation has led to several initiatives to improve the governance of Earth orbital space, which are discussed below. These initiatives, if successful, will in time contribute to a consensus on the fair and responsible use of outer space and also to the ability to maintain sustainable space activities.

Over the past decade the world has seen a sharp increase of space-faring States, those who own and operate their own satellite, from 27 to 37. This number is expected to increase steadily in the future. Further, the number of States capable of launching their own spacecraft into orbit is also increasing, to a total of eight.[2] In February 2009, for example, Iran became the latest entrant to the launch-capable club of States by launching a small experimental satellite on an indigenous launcher.[3] Increasing numbers of spacecraft in Earth orbit means that some orbits, specifically the geosynchronous orbits (GSO) ideal for communications and the sun synchronous orbits (SSO) favoured for Earth observing satellites, are becoming more crowded, which increases the probability of accidental spacecraft–spacecraft collisions in these well-used orbits.

The most likely collision possibilities are between working satellites and space debris. The U.S. Space Track Catalogue contains some 13,000 orbital objects greater than 10 cm in extent, of which about 900 are working satellites. The remainder is space junk. The U.S. Air Force tracks an additional 6000 or so orbital objects that do not appear in the catalogue because their launch origins are not known. Above about 400 km, Earth's atmosphere exerts very little drag on orbital objects, resulting in longer and longer times before they decay from orbit. Thus, objects in very high altitudes, above 800 km or so, will stay there for centuries, continuing to threaten working satellites. Worse yet, these objects move around over time, perturbed by pressure from the sun and gravitational forces, thus requiring constant vigilance to avoid. The world experienced an unfortunate example of this possibility on 10 February 2009 when a low-Earth orbit (LEO) commercially operated Iridium communications satellite collided with a non-functional Russian Cosmos satellite, leaving two more debris clouds to threaten other satellites.[4] Extensive media coverage of this incident has led to heightened awareness among the general public as well as the professional space community of the need for greater governance of outer space.

The following initiatives, which are discussed in more detail in other articles in this volume, contribute to the development of consensus on what constitutes the fair and responsible use of outer space.

2.2. Continuation of States' efforts to develop robust methods to curb the production of new orbital debris, in accordance with the UN guidelines on orbital debris[5]

The development and passage of these guidelines, which were derived from the work of the Inter-Agency Debris Coordinating Committee (IADC), was an important step towards reducing the generation of debris that would add to the already serious problem faced by operators of spacecraft orbiting Earth, but it was just a first step in the process. Although these debris-reducing efforts are important in limiting the production of new debris placed in orbit from space operations, they are hardly enough. For these guidelines to be effective, the space-faring community must strengthen them by incorporating the guidelines firmly into State operating practices, regulations and other mechanisms for implementation. Fortunately, States are making progress in this regard. The meeting of the 2009 UN Legal Subcommittee of COPUOS revealed that several States have indeed made progress in that arena, which has strengthened interest in broader State compliance.[6] Nevertheless, additional progress on extending these rules must continue in order to ensure that more States institute firm debris controls.

2.3. Acceptance and promulgation of a set of best operational practices in space

Led by France, States and commercial entities have entered into an exercise to identify and codify a set of "best practices" for States and commercial entities to follow when conducting space activities in Earth orbit. Such practices include limiting the generation of orbital debris, but extend further to include separating transmission frequencies in order to avoid interference with other operators, safety of space operations, monitoring of space weather and other operations-related practices.

2.4. Efforts towards developing a Code of Conduct or rules of the road for space operations

The "commons" nature of Earth orbit has become ever more evident in recent years and several entities have endeavoured to develop a code of conduct, or rules of the road, for States to follow in conducting space activities. As if to underscore the nature of the space commons, the collision of the Iridium-33 commercial communications satellite with the inoperable Russian Cosmos-2251 communications satellite on 10 February 2009 produced debris that soon after threatened the Canadian Radarsats I and II.[7] As discussed elsewhere in this book, in 2008, the European Union took on the task of developing a draft Code of Conduct for consideration by the international community. The EU's goal is to have countries agree to a draft in the near future in order to improve safety and security in Earth orbit.[8] It has been consulting with other space-faring States in order to develop a commonly agreed code of conduct that all can follow in the future.

2.5. Space situational awareness (SSA)

In order to prevent further collisions in Earth orbit, spacecraft operators will have to have much better awareness of the potential threats they face from other spacecraft and from orbital debris. This includes not only tracking the 19,000 or so total objects in orbit, but also continually calculating the chances of another accidental collision in order to avoid creating additional debris. Unfortunately, neither effort is a trivial exercise and requires considerable tracking resources (optical telescopes and radar), computing power, and sophisticated software to calculate the thousands of possible daily satellite–satellite or satellite–debris conjunctions. Currently, only the United States has anything close to the global network of observational resources and computational capability necessary to carry out this task, though not to support the entire computational effort. However, Russia, China, France, Germany, Norway and the European Space Agency also have some tracking and computational capacity. In addition, the International

Space Observation Network (ISON), an international network of astronomical observatories, collects high accuracy data on objects in GSO. Finally, several GSO communications satellite operators are beginning to share some details of their positions with each other through a trusted data centre partnership. The satellite operators have the most accurate positional data on their own satellites and the combined data set adds additional space situational information, at least for GSO.

Although the United States shares its space object catalogue at www.space-track.org with any individual or any organisation that wishes to use the data, the open catalogue is derived from the much more accurate Air Force catalogue, the data of which are classified, in large part because they contain other details about the objects, including function and estimated operational status. Nevertheless, the positional data are of most interest to the commercial and civil government community of satellite operators. Hence, if all of the accurate positional data from the U.S. catalogue could be made public and integrated with data from both the ISON network and data collected by other countries, satellite operators could do their own conjunction analysis, a task that is virtually impossible today because the data available publicly are too limited or lack sufficient accuracy. Secure World Foundation believes that the United States would do well to open the accurate positional catalogue to the world and urge other States to do the same. This would enable the relatively rapid development of an international cooperative SSA effort in support of commercial and civil government needs for space situational awareness and conjunction analysis, leading to a much more secure space environment.

2.6. Space traffic management (STM)

As orbital space becomes more crowded with debris and with functioning satellites, efforts beyond an international civil SSA may be needed. In addition to just tracking satellites, it may be necessary to regulate and manage their operations. This concept has been explored in detail in reports of the International Academy of Astronautics[9] but will require extensive study by the space-faring States and extensive coordination to make such a major change in their operational approaches to space activities. Experts still disagree over whether an STM regime will ever be needed or desirable, yet in my view, some form of STM is inevitable and the international community should be exploring today the benefits and drawbacks of developing and instituting such a system.

2.7. Banning debris-creating space weapons

Of potentially far greater danger to the continued viability of the space environment than debris from normal space operations is the development, testing and use of debris-causing space weapons. The deliberate destruction of a defunct Chinese Fengyun 1-C meteorological satellite by a People's Liberation Army (PLA) ballistic missile in January 2007 generated more than 2300 pieces of trackable debris, which have since threatened other satellites.[10] This is the largest known amount of trackable orbital debris to have been generated in Earth orbit from a single event. Many tens of thousands of much smaller debris were undoubtedly created at the same time but cannot be tracked because they are too small for current tracking methods. Yet, this debris also poses a serious threat to space operations, especially those in SSO, because at orbital speeds, even very small objects can severely damage or destroy a spacecraft.

Clearly, the era of treating orbital space as a realm in which there are few rules beyond the stringent laws of physics is over. The question for space-faring States is now what sort of rules are needed and how they should bind States to make sure that activities in outer space are conducted in a fair, responsible and sustainable manner. Taken together, the initiatives summarised above can support the development of the necessary consensus to enhance the regulation of outer space.

UNCOPUOS is the primary international entity for arriving at a global consensus on what constitutes the fair and responsible use of outer space. For many years, COPUOS has seemed to be absent from this important role. In recent years, however, I sense a new vibrancy in the Committee and a new willingness on the part of COPUOS delegates to tackle some of these important issues and contribute to the development of a new order in space, one where States actively contribute to the long-term sustainability of space activities.

Although there is still time to realise the initiatives summarised in this article, the international community needs to pursue them with vigour lest the commons of Earth orbit become an impassable junkyard and a testament to humanity's inability to work sustainably in Earth orbit.

---

[2] China, France (ESA launcher), India, Iran, Israel, Japan, Russia, United States. The European Space Agency operates its launch facility in French Guiana.

[3] Tait, Robert. "Iran Launches First Domestically Produced Satellite." 3 Feb. 2009. guardian.co.uk. 15 June 2009. http://www.guardian.co.uk/world/2009/feb/03/iran-satellite-launch-omid; "Jonathan's Space Report" No. 606, 16 Feb. 2009. 15 June 2009. http://host.planet4589.org/space/jsr/back/news.606.

[4] Weeden, Brian. "Billiards in Space." 23 Feb. 2009. The Space Review. 15 June 2009. http://www.thespacereview.com/article/1314/1.

[5] "Space Debris Mitigation Guidelines of the Scientific and Technical Subcommittee of the Committee on the Peaceful Uses of Outer Space." 6 Mar. 2007. Report of the Scientific and Technical Subcommittee on its Forty-fourth Session, Held in Vienna from 12 to 23 February 2007. General Assembly Document A/AC.105/890. 15 June 2009. http://www.oosa.unvienna.org/pdf/reports/ac105/AC105_890E.pdf.
[6] Legal Subcommittee 2009: Forty-eight Session (23 March–3 April). United Nations Office for Outer Space Affairs. 15 June 2009. http://www.oosa.unvienna.org/oosa/en/COPUOS/Legal/2009/index.html.
[7] Haynes, Megan. "Space Junk Endangers $1B Canadian Satellites." 21 Mar. 2009. Canada.com. 15 June 2009. http://www.canada.com/Technology/story.html?id=1414605; "Duo of Canadian Sats in Debris Danger." 23 Mar. 2009. Satnews.com. 15 June 2009. http://www.satnews.com/cgi-bin/story.cgi?number=1849002796.
[8] "Draft Council Conclusions Concerning the Draft Code of Conduct for Outer Space Activities and Draft Code of Conduct for Outer Space Activities." 1650/08. 3 Dec. 2008. Council of the European Union. 15 June 2009. http://www.eu2008.fr/webdav/site/PFUE/shared/import/1209_CAGRE_resultats/Code%20of%20Conduct%20for%20outer%20space%20activities_EN.pdf.
[9] Contant-Jorgenson, Corinne; Lála, Petr and Kai-Uwe Schrogl, eds. Cosmic Study on Space Traffic Management. Paris: International Academy of Astronautics, 2006.
[10] "Fengyun-1 C Debris: Two Years Later." Jan. 2009. Orbital Debris Quarterly News 13. 1. 15 June 2009. http://www.orbitaldebris.jsc.nasa.gov/newsletter/newsletter.html.

3 United Nations' space policy to preserve peace and sustainable development

Ciro Arévalo-Yepes

Fifty years of contemporary space history demonstrate that uses of space and its natural resources serve mankind's critical needs and interests. We are facing a growing awareness of global interdependency in which worldwide approaches to economical issues are necessary to face major global crises. The UN's challenge is to assume the leadership needed to respond in a fair and responsible manner to this unprecedented situation. Transforming their unequivocal diagnostic platforms into an operational delivery process suited to societal needs seems to be a central institutional objective.

At a recent high-level event on the food and climate change crises, held in New York on 25th September, UN Secretary-General Ban Ki-Moon identified the ongoing global food crisis, together with the climate change and the natural disasters, as three of the key challenges of both our current generation and the generations to come.

United Nations data suggest that an additional 75 million people have slid into hunger due to the food crisis, lifting the total figure now to well above 900 million, and the situation may indeed get worse as the world's population is expected to grow by one-third over the next 40 years. World food demand will double within that same time frame, water insecurity in all parts of the world will increase and so will the effects of land degradation and climate change.

We also remember the devastating cyclone that hit Myanmar earlier this year, and the earthquake that claimed thousands of lives in China. These examples demonstrate once again how vulnerable we are to the forces of nature, and how important it is to improve our capacity to mitigate the devastating effects of disasters. Loss of life and property could be avoided if better information were available through improved risk assessment, early warning and better monitoring of disasters.

To be sure, we are thus dealing with an area of fundamental importance to our current and future life on Earth. There are truly inter-linkages between food security-related areas and other areas of major concern, such as climate change, health, energy, water and disaster management. We need to address these issues in a holistic manner and look into a variety of tools for solutions. Space technology applications provide one such tool-set of increasing importance in decision-making processes at all levels.

The international community observes how the UN assumes its global responsibility in order to face those challenges. Responding to this call, COPUOS reinforces the current means and opens new venues to coordinate the efforts made by the UN in order to cope with the whole situation. The Committee has made substantial efforts to identify synergies with the results of global conferences held within the United Nations system and other global initiatives, particularly with the United Nations Millennium Declaration, the plan of Implementation of the World Summit on Sustainable Development, the plan of Action of the World Summit on the Information Society among others.

The fair and responsible uses of space also require a new vision about how space activities could be technically and equitably sustainable in time and scope. Competition and cooperation have to be pondered upon and must be combined if we want to preserve the interests of our ultimate constituency: the community of users. The growth of an impressive number of space actors from emerging space-faring nations, private, institutional and intergovernmental shareholders has given birth to an intricate network of users and providers that requires a good knowledge of the international legal framework in which they should operate but also deserves an innovative set of guidelines for situations that are new.

The scientific evidence that we have reached a state of saturation of some key orbits, along with the increasing amount of orbital debris and the preservation of peaceful uses of outer space, poses central challenges to the United Nations as a whole. In order to create a culture of space security, innovative multiform cooperative structures are required. In that regard, a few proposals are emerging today aimed at a long-term sustainability of space activities such as a code of conduct for space activities and space traffic management. In my opinion any new initiative that is expected to reach a basic universal consensus has to satisfy tree basic conditions: (1) it must take into account the existing legal instruments; (2) it must include a consultations process between major space-faring nations and developing countries and (3) it must contain a multilateral framework with a long-term perspective.

It is necessary to recall that present space achievements are not the result of spontaneous generation, but a transversal knowledge accumulation with global contributions from all main world geographical regions. The map of Space activities today is very impressive. At the end of the second world war, the total number of countries was 50; today almost 200 countries are members of the United Nations. To have their own satellite has become for many of them a matter of national pride. It is estimated that there are around 50 countries that have some kind of participation in the fabrication of a space object while some regions of the world have an impressive evolution in this area. In the last few years, for example, five Latin American countries have created some type of national space coordination bodies. In other regions of the world the evolutions have been outstanding too. Having representatives of all geographical regions, COPUOS has been the initiator and promoter of many of the space cooperation agreements signed by its members. Their respective interaction gives the Committee vitality and commitment.

The need now, as never before, is to achieve a common understanding regarding the meaning of "fair and responsible uses" and "peaceful purposes" which are three strong interrelated concepts that must be analysed in a holistic way. We have continued to build a common understanding of those major concepts from different perspectives. As regards the concept of a fair and responsible use of outer space itself, we already have at hand some accepted and relevant principles consecrated in the UN charter and in the main legal instruments that compose international space law.

3.1. Contributions of COPUOS relating to this topic

COPUOS and its Legal and Scientific and Technical Subcommittees have provided critical institutional leadership in the development of the main space legal and cooperation processes. The Province of all Mankind Concept, for instance, has been considered as a means for fairer and more orderly set of human relationships and, by definition, a more responsible way to assume them. This is probably the strongest unifying concept since it ensures that the space environment be extended and used for the benefit of all States regardless of their economic and technological level. The mankind concept reapplies the true sense of the *res communis* principle to outer space – the moon and celestial bodies. In my opinion, we are not fully aware of the impact of this principle and it's necessary to call it up in whichever activity is undertaken today.

Another important COPUOS contribution in that regard concerns the 1986 remote sensing principles, which demonstrated that a consensus can be achieved among a group of States with strong asymmetries in their technological capacities and in spite of diverse political motivations. Regarding those remote sensing principles, the developing countries managed during the negotiations to obtain 8 of the 15 principles that represented substantial benefits for them. Nevertheless, many efforts need to be made, particularly on the cost of and access to satellite images, even though some mechanisms were established such as the UN Charter of Major Disasters and recently UN-SPIDER.

Other resolutions adopted by the General Assembly were based on a fair and responsible perspective of the use of outer space. We have paragraph 4 of resolution

55–122 of 8 December 2000 relating to the rational and equitable use of geostationary orbits. The Legal Subcommittee recommended a coordinating mechanism to access the spectral orbit resources in order to guarantee an effective use of the orbit/spectrum taking into account the needs of developing countries. Despite what still needs to be done, particularly in the ITU context, this contribution by COPUOS represented the basis of a more rational, efficient and economic use of the GEO and complemented the application of article 44 of the ITU Constitution as amended by the Plenipotentiary Conference held in Minneapolis in 1988.

In the area of space applications, COPUOS has continuously made efforts to promote and increase awareness and capacity-building in the use of space technology applications, at the international, regional and national level, in many critical areas of concern for all mankind. Space technology and its applications, such as earth observation systems, meteorological satellites, satellite communications and satellite navigation and positioning systems, provide effective tools for monitoring and conducting assessments of the environment, handling its natural resources, providing early warnings of disasters and ways to manage them, as well as providing education and health services in rural and remote areas.

The Committee has been responsible for organising three United Nations conferences on the exploration and peaceful uses of outer space. With the organisation of the already one decade old UNISPACE III in 1999, and through its implementation, the Committee has aligned many of its activities with the major UN main forums. The results are quite impressive and it is with good reason that we will celebrate its 10th anniversary next year.

In that regard, I would like to recall the unique approach of the Committee to implement the recommendations of UNISPACE III through the establishment of action teams in priority areas. The establishment of UN-SPIDER under the Office for Outer Space Affairs and the establishment of the international Committee on Global Navigation Satellite Systems (ICG), which provides a forum for coordination and cooperation among GNSS providers and users, are the two outstanding examples of those concrete results.

The Committee is convinced that international space law without reference to critical human values is a sterile regime. Therefore it has been instrumental in the creation of a legal framework governing the activities of States in the exploration and use of outer space, consisting of five treaties and five sets of declarations and principles on outer space activities. Among them, the Outer Space Treaty of 1967 represents a landmark legal instrument – the "Magna Carta" of space law – whose 40th anniversary we celebrated last year. The Outer Space Treaty, together with the other core treaties on outer space, forms the legal order for today's space activities. It is with great pleasure that I today observe the efforts of the Committee and its Legal Subcommittee to further advance the application of the legal regime of outer space and to promote capacity-building in space law.

In that regard, the Committee and its subsidiary bodies made considerable achievements last year with the adoption by the General Assembly in its resolution 62/217 of the space Debris Mitigation Guidelines of the Committee on the Peaceful Uses of Outer Space and the adoption of resolution 62/101 on registration practice. As far as the space debris mitigation guidelines are concerned, I have received strong suggestions that we should have a more robust type of regime and not only an optional one.

Both Subcommittees could start considering two new agenda items on capacity-building in space law and general exchange of information on national legislation relevant to the peaceful exploration and use of outer space. These achievements demonstrate once again a considerable level of activity by the Committee and its subsidiary bodies in areas of utmost importance to current and future activities in outer space.

As far as the nuclear power sources in outer space are concerned, the Committee has established a new partnership with the International Atomic Energy Agency (IAEA) to prepare a safety framework for nuclear power source applications in outer space. COPUOS is also concerned with the threat of the near-Earth objects and is considering a range of suitable mechanisms for handling this issue at the international level.

The Committee entered a new chapter in this historical path by further developing in an active way its overall agenda aimed at meeting development needs. The Committee this year agreed to include in its agenda for 2009 two new items: (1) space and climate change and c(2) the use of space technology in the United Nations system. The committee is continuously considering items regarding inter alia space and water; international cooperation in promoting the use of space-derived geospatial data for sustainable development; and space and society. The Committee is also actively increasing its efforts in contributing to the thematic clusters under the work programme of the Commission on Sustainable Development.

In this regard, I would like to emphasise the role of the Inter-Agency Meeting on Outer Space Activities that restructured its reporting mechanism and strengthened its important role vis-à-vis the Committee. The revitalised reporting structure and the means provided by the new agenda item will pave the way for a fruitful and mutually beneficial role of the Inter-Agency mechanism and the Committee, and we expect to have a stronger Member State participation in its working.

As mandated by the Committee last year, the Office for Outer Space Affairs, together with an expert group of educators in space law and the Directors of the

Regional Centres for space technology education, affiliated with the United Nations, started in December last year to elaborate and develop curricula for space law education for the Regional Centres. This project is particularly interesting since such curricula could eventually also be used for teaching space law at universities and academic institutions around the world.

3.2. Visions and personal opinion

As Chairman of the Committee for the 2-year period form June 2008 to June 2010, I would like to promote the vision that the Committee will continue to strengthen its major role in shaping the international standards for space activities and in promoting international cooperation at the global, regional and inter-regional level for the benefit of all countries. In this regard, the United Nations family, through its entities and programmes, plays a fundamental role in contributing to the advancement of such endeavour and to adjust it to a new space environment. The United Nations, in coordination with its Member States, faces an important challenge, which is to use non-traditional tools like those related to space technology in order to deliver better results and, subsequently, a more concrete approach to global issues. In that regard, in my personal view, it is time to think about a celebration of a UNISPACE IV.

It is also of fundamental importance for COPUOS to gain strong visibility not only within the UN system but also with its specialised agencies, as well as other organisations, particularly space agencies of advanced space-faring nations and to improve its relationship with the space coordinating bodies of developing countries. Within the Committee itself, a better monitoring of the contributions by its permanent observers is also necessary. Structural projects have to be reinforced with key partners like UNESCO, ITU and IAF among others and to share COPUOS main results with the Conference on Disarmament.

It is also necessary for COPUOS to have a strong Secretariat. I am confident that the Director of OOSA, Mrs. Mazlan Othman, will reinforce the Secretariat's capabilities in order to implement COPUOS decisions, particularly those dealing with technical and legal cooperation with its Member States.

Space applications are multifaceted and often offer, through a single instrument or application, the means for States to make development decisions concerning various crosscutting issues and challenges. Those instruments facilitate the implementation of actions called for at the global, regional and inter-regional level. The reinforcement of regional cooperative structures in Africa, Asia-Pacific and Latin American regions will be one of the priorities complementing the north–south cooperation. The creation of a Latino American Space Policy Center for example is a good initiative in that direction. As it has been expressed before me, I am convinced that international cooperation is the best tool for maintaining the peaceful uses of outer space.

Finally, the design of a road map for the United Nations space policy should become an important objective for the years to come. There is a continued need to strengthen unified efforts at all levels and among all relevant stakeholders in addressing overarching long-term development concerns and to further enhance a sure and equitable use of space for the benefit of all humanity. I am confident that we will make important progress in this direction in the coming years.

4 Current questions regarding fair and responsible use of space – summarising introduction

Niklas Hedman

The first session of the Workshop reviewed overarching topics on the role of States and other actors in meeting common challenges to sustainable space governance. In determining the meaning of fair and responsible use of outer space in its broader perspective, the session addressed both civil and military use of outer space and focused its attention on the overall role of space-faring nations and emerging space nations in advancing international cooperation in space activities. The session comprised the following speakers: Peter Martinez on Fair and responsible uses of space: a perspective from an emerging space country"; Theresa Hitchens on "Peaceful use of outer space vs. militarisation: cost–benefit analysis"; Fernand Alby on "The space debris environment and its impacts"; and Giovanni Gasparini on "Space Situational Awareness: an Overview".

Peter Martinez in his presentation pointed to the large number of countries that lacked means to take full advantage of the application of space science and technology, and stressed the need for stronger involvement of all countries in efforts to enhance international cooperation in space activities. Theresa Hitchens made an analysis of military, economic and social benefits of space activities, and expressed that, given the increased population of space debris, which posed a real threat to both military and civil satellites, including commercial satellites and consequently to commercial interests, and the risk of conflict escalation and increased tension, there was indeed a high price to pay for "weaponisation" of outer space. Fernand Alby, who made an analysis of the Space Debris Mitigation Guidelines of the Committee on the Peaceful Uses of Outer Space in comparison with the debris mitigation guidelines of the IADC and other mechanisms, presented the view that in order to prevent future creation of space debris in orbit, there was a need for a stronger international regulatory regime with compulsory mechanisms. Giovanni Gasparini addressed Space Situational Awareness (SSA) as mechanisms to provide better knowledge and understanding of space assets and space-based systems in orbit, and the potential role such mechanisms, predominantly security-driven tools, could play for future information sharing and confidence building.

The discussions, following the four presentations, focused to a large extent on the role of emerging space nations and developing countries in current and future international cooperation. In view of changes in the geopolitical landscape over the past 50 years and the increase of actors in the space field, including from private and commercial sector, the need to better integrate different stakeholders, both governmental and non-governmental, and to better promote the access to space-based data was acknowledged. A need for a holistic approach in terms of protecting the Earth environment and space environment was determined, and through international cooperation there was a need for achieving better fairness and responsibility in space activities. The common use of outer space for peaceful purposes and the avoidance of allowing outer space become a theatre of war and conflict was emphasised. The strengthening of existing legal regimes and the need for enhanced application and implementation of key legal instruments were mentioned as important factors to achieve fair and responsible conduct of space activities. Likewise, in view of greater knowledge and awareness of the fragility of the space environment and the need for enhanced sustainability, there was a need to develop further mechanisms and regulatory frameworks for the protection of the space and planetary environment. The usefulness of looking more closely into different tools for the management of space activities, such as SSA mechanisms, and to strive for broader participation and application involving developing countries was also stressed.

4.1 Fair and responsible uses of space: a perspective from an emerging space country

Peter Martinez

Abstract

This paper examines the different notions of "fair and responsible" uses of space from the perspective of advanced and emerging space nations. For emerging space nations, access to space applications to support human and environmental security constitutes an important element in their understanding of "fair and responsible" uses of space. Space applications also support social development though provision of services and through supporting improved governance. The rapid evolution of the space arena means that there are now many more actors and many more possibilities for cooperation open to emerging space nations than in earlier years. This, too, raises issues of the "fair and responsible" use of space for both advanced and emerging users of the space environment. The paper ends with a brief reflection on whether the notion of "fair and responsible" use of space applies to space exploration from the perspective of developing countries.

4.1.1. Introduction

If one were to ask a number of people what they understood by the term "fair and responsible use of space", the answers would probably depend on the national origins of those answering. A person from an advanced space-faring nation might understand this term in the context of "space security" (which is in itself a term open to various interpretations), but essentially pertaining to issues *in space*. A person from a developing country without a space programme would have a very different view. To that person, "fair and responsible use of space" would pertain to issues *on the ground*, such as access to data or technology. The aim of this paper is to consider these two complementary sets of views in light of the rapidly changing international space arena.

When India established its national space programme in the early 1970s, it was with a clear view to using the unique perspective of space to address problems of national development, as so clearly elucidated in the following oft-quoted remarks by Vikram Sarabhai, the father of the Indian space programme:

> *"There are those who question the relevance of space activities in a developing nation. To us, there is no ambiguity of purpose. We do not have the fantasy of competing with the economically advanced nations in the exploration of the moon or the planets or manned spacecraft. But we are convinced that if we are to play a meaningful role nationally and with the community of nations, we must be second to none in the application of advanced technologies to the problems of man and society, which we find in our country."*

Sarabhai's comments were right on the mark. Today, space applications affect humanity and society in a major way, and among developing nations, India is a leader in the applications of space technology to societal development. Indeed, it could be said that space applications are now part of the plumbing of modern life, even in many developing countries without space programmes. So, what does the concept of "fair and responsible use of space" mean for emerging space countries, or, indeed, for countries that do not have space programmes?

In order to answer this question, we need to consider the particular needs and challenges facing developing countries in applying space technology, as well as the changing geopolitical context of space activities. This context today is very different (and changing rapidly) from what it was at the beginning of the Space Age. This different, changing context dictates that emerging space nations will follow a different trajectory of space development than that followed by the first space powers.

4.1.2. Space applications and the developing world

It is estimated that the world's population will reach about 9 billion people by the year 2050.[11] Almost all the population growth from the current 6.75 billion will take place in less developed countries. The growing population is placing growing pressure on the environment to meet adequately *all* the resource needs of *all* the Earth's inhabitants. This is a global challenge, not just one for developing nations, and space has the global reach to address it. Space applications support environmental security, disaster management and human security and form one of the cornerstones of the Information Society.

The space contributions to these areas are reviewed briefly in the following sections, with emphasis on their particular importance in the developing world.

4.1.2.1. Environmental security

It is a truism that the planet that matters most to us is the one we live on. This fact predicates the logics of the space programmes or space applications capabilities of developing nations. It also shapes their perceptions of what constitutes "fair and responsible" use of space.

Among a host of environmental issues, global climate change is currently a major political and scientific concern worldwide. Developing countries are particularly vulnerable to the effects of climate change. Space technology has played a significant role in establishing the scientific basis for determining that this is a real issue requiring collective and long-term action by the community of nations. There is now sufficient consensus that action is required, leading Governments to allocate resources to address these challenges and to enter into treaties that address policy areas of mutual concern.

Earth observation satellites support environmental security by providing global observations of many important environmental parameters. However, there is a need to address issues such as the lack of data consistency, fragmentation, gaps in coverage, as well as data accessibility and interoperability challenges. Programmes such as the Global Earth Observation System of Systems (GEOSS) and Europe's Global Monitoring for Environment and Security (GMES) have started to address these data-related challenges.

There is another way that Earth observation satellites support environmental security, and that is through supporting the development and implementation of environmental treaties. This occurs throughout the whole treaty cycle. In the pre-negotiation phase, satellites can be used as an agenda-setting aid by providing global data that help us to identify and quantify the extent and magnitude of the problem. In the negotiation phase, satellite data can be used for assessments of the status quo and for target setting. In the implementation phase, satellites support monitoring and reporting requirements (e.g., as stipulated in the Kyoto Protocol). Satellite data can support treaty enforcement at national and international level in matters such as compliance verification and dispute resolution. A more detailed consideration of the role of satellites in supporting environmental treaties is given by Peter.[12]

Much work has been done by many actors to promote the usage of space applications for development. Despite this, there are still a large number of countries that lack the means to take full advantage of space applications benefits for environmental security. There are several reasons for this:

(a) *Earth observation applications on environmental issues are in the nature of a "public good" activity.* Such activities are normally undertaken by governments, but governments in developing countries often lack the resources and expertise to tackle the issue. Since there is no commercial incentive to develop applications locally, industry does not play a role, and hence the potential benefits are not realised.
(b) *Affordable, timely, appropriate and complete data are often not available.* The cost of data access can be prohibitive for developing countries, or the necessary data are simply not available.
(c) *The challenge of operationalisation.* The space community is replete with examples of successful pilot projects by space agencies that demonstrate the utility of space applications to environmental issues. However, it is very difficult to convert these projects into *operational* programmes.

With regard to (a) and (b), initiatives such as GEOSS and GMES are starting to address some of these issues. With regard to (c), I believe that there needs to be a closer engagement with the development sector to get space into the mainstream of development aid thinking. Often space is regarded as "too high-tech" for developing nations and is not seen as a development tool, in the same way as providing pumps or trucks, or building roads. Space is part of modern infrastructure for enabling the information society, just as the traditional infrastructure of roads, railways and harbours enabled the industrial society. Seen in this light, the development aid sector might be prepared to work more closely with the space sector to help build up the modern infrastructure of developing countries. This could involve the private sectors in developed and developing countries, thus addressing issue (a) above as well. Indeed, such an approach could contribute to a more "fair and responsible" use of space by the developed space-faring nations to address global environmental security concerns to the benefit of all nations. In so doing, it would de facto create more equitable access to space benefits for all nations.

4.1.2.2. Space applications for disaster management

If there is one area where the different types of space applications complement each other perfectly, then it is in the area of disaster management. Space applications can be used in all phases of the disaster management cycle – from risk assessment, through the acute disaster phase and the recovery phase afterwards. Because of the humanitarian aspect of disaster management, this is perhaps the area with the best examples of equitable access to the benefits of space.

The challenge to successful implementation of space-based disaster management can be summarised as the need to deliver:

- the right information;
- in the right form;
- to the right people;
- in the right place;
- at the right time.

Space tools can address each of the links in this chain, but in practice this is a complex challenge to overcome. The reason for this is that it is not the technology per se that is the problem, but rather the interface between the space community, who are comfortable with space tools, and the disaster management community, who are not familiar with these tools. In order to make full use of space-based disaster management tools, ways must be found to integrate space more effectively into intuitive, easy-to-use tools for responders on the ground. This could be accomplished through exploiting existing interfaces known widely to non-space users (e.g., Google Earth). Responders also require training in the use of even "simple" space tools. For example, responders were issued with satellite phones during the hurricane Katrina disaster. Many problems were experienced by first-time users unfamiliar with this technology; e.g., trying to make calls from locations where no satellite signals could be received, or thinking that the satellite phones could overcome disrupted terrestrial mobile services, and so on.

The International Charter on Space and Major Disasters has proven to be an extremely positive step for securing urgently required imagery of areas affected by disasters. Since January 2002, the Charter has been triggered 183 times for disasters on all populated continents. In spite of this, some nations still report "difficulty" in activating the Charter. This points to problems of coordination: the Charter coordinates capability in space, but it seems that coordination is lacking on the ground in the disaster management community in some countries. Simply put, the mechanism for triggering the Charter in those countries is not yet well known and established in their disaster management communities. This is perhaps one area where the UN-SPIDER initiative for coordinating space-based disaster management will be able to make a difference.

Instruments such as Charter and the new UN-SPIDER platform for space-based disaster management are examples of "fair and responsible" uses of space for the benefit of all mankind.

4.1.2.3. Space and social development

Satellite communication is a cornerstone of the global information society. It is also a potent enabler of development, especially in countries with poor fixed terrestrial communications infrastructure. The following are some of the activities enabled:

(a) *Tele-medicine/health and tele-education:* Tele-health services can help developing nations to inform their populations on a wide range of health issues (such as HIV/AIDS, avoidance of diseases such a cholera, etc.) and to overcome shortages of doctors in remote areas. Tele-education provides the means to overcome shortages of teachers and teaching materials in remote areas.

However, four elements all need to be in place for a successful operational implementation of such services. First, connectivity must be robust. The infrastructure and support system must be established (ground terminals, training, technical support). Second, the tele-education/health activities must be integrated with the local educational system (and curriculum!) and with the local medical services, respectively. Third, educators and medical personnel must be properly trained in the use of this technology. Fourth, there should be seamless integration of the distant and in situ health/education systems. These are all issues that must be resolved *on the ground*, not in space. Among developing nations, India provides perhaps the best examples of space systems that are fully integrated into a variety of areas, ranging from fisheries to agriculture, medicine and education.

(b) *Improved governance:* Satellite communication can also be used to "bring government to the people". For example, people in rural communities without government offices can access government information and submit forms electronically ("e-filing") without the expense of travelling to cities to submit applications for identity documents, birth certificates, pension payments, etc. This technology has also been used in South Africa to link the people in remote isolated communities to their elected representatives in Parliament, over a 1000 km away.

With regard to "fair and responsible" use of space in the domain of satellite communications, the dominant issue for the developing countries is access to spectrum and orbital slots in geostationary orbit. Every year, developing countries

Fig. 1. *2007 World Radiocommunication Conference (source: ITU).*

in COPUOS reiterate this view and point out that there is a difference between the ideal "paper" situation and the situation in practice in geostationary orbit.

In the developing world the C band is widely used, partly because there is a well-established infrastructure in place, but also because of the cloud-penetrating ability of C-band radio frequencies. But there is growing pressure on the C band from military and civilian users. In the 2007 World Radiocommunication Conference the satellite communications operators led a strong lobby against a push by the terrestrial mobile technology operators to use the C band. The Congress voted to safeguard the C band for satellite users, essentially affirming the view that this is a "fair and responsible" use of space (Figure 1).

4.1.3. The evolving space arena

4.1.3.1. From prestige-driven space age to information-driven space age

The space arena was shaped historically by the cold war context in which it developed and was driven by rivalry (political, military, economic) between the two key actors, the USA and the USSR. The principal driver was international prestige for their two competing economic and political systems. The early Space Age was marked by competition, rather than cooperation, and space was seen as a platform for projecting national power. Cooperation in space activities was therefore mostly intra-bloc, with a few notable exceptions, like the Apollo-Soyuz Test Project.

We are now in a second Space Age, which is driven by *information* as a commodity. This period began with the end of the Cold War and is characterised by the use of space technology to provide information for security and prosperity. This period has seen the emergence of the commercial space sector as a major player, both in terms of the space industry and terrestrial industries based on assets in space. The former includes the large space contractors and the many smaller companies that form part of the global space system supply chains. The latter includes the communications and broadcast satellite operators, the GNSS industry and an emerging Earth observation data distribution industry (along the lines of Google Earth). Together, these industrial groups have allowed the emergence of global utilities, such as satellite navigation/timekeeping, communications and Earth observation, bringing the benefits of space applications directly to many millions of individual users.

4.1.3.2. Growth in number and diversity of actors

There are now many more actors in the space arena than in the first two decades of the Space Age. By 2005, there were 36 national space agencies (on all continents).[13] By 2007, 10 actors had demonstrated independent orbital launch capability and 47 States had launched civilian satellites, either independently or in cooperation with others.[14]

In the early days of the Space Age, the actors in the space arena were all States and their national space agencies. However, there is now a much more diverse set of actors. Industry has become a major player in terms of enabling new actors to enter the arena. Non-governmental organisations (NGOs) are also playing a significant role in allowing new actors to enter the arena. This is especially so in the case of international professional organisations, such as the IAF, the IAA, COSPAR and others. These NGOs provide a forum to bring together a diverse set of actors in an informal setting to discuss matters of mutual interest in a way that may not be possible in inter-governmental organisations.

The appearance of more actors and a higher level of activity in space means that there will be more pressure on available orbital and spectrum resources, more pressure on the space environment and consequently a greater need for coordination. On the other hand, it also means easier access to space data and services from a variety of sources. Established and emerging space nations and non-space nations will all have different views on what constitutes "fair and responsible" uses of outer space. These matters will have to be debated in global space fora to reach a common understanding and to define accepted rules of conduct.

4.1.3.3. Increasing reliance on space capability by the military in more countries

An increasing number of States are making military use of the full range of space applications. To date, 14 States have launched dedicated military satellites and as more countries enter the space arena, it is likely that their militaries will follow. The military also makes use of commercially available data (e.g., Earth observation) and services (e.g., communications) to supplement their own capabilities.

There is also a trend to integrate military and civilian applications into dual-use satellites. This makes sense from a technological perspective since there is no intrinsic difference in the technologies required by civilian and military users. The only difference is in the applications. Such dual-use systems are also attractive for emerging space nations, which may not have the resources to develop separate military and civilian space programmes.

However, there are also some disadvantages to combining civilian and military functions on a single satellite. Dual-use satellites can become potential targets in conflict situations and they could also be an impediment to cooperation or commercialisation if the military partner is sensitive about the technical details or orbital parameters becoming widely known.

4.1.3.4. Changing patterns of cooperation

In the early days of the Space Age, international cooperation was mostly intra-bloc, with few exceptions. With the growth in the number of space-faring countries, there is now a much wider spectrum of possibilities for cooperation.

A number of regional cooperation structures have emerged over the past 20 years. These structures have arisen from initiatives by the leading space countries in each region, acting as aggregators to promote the application of space technology throughout their region. In the Asia-Pacific region, two regional structures have emerged, the Asia-Pacific Space Cooperation Organisation (APSCO), under the leadership of China, and the Asia-Pacific Regional Space Agency Forum (APR-SAF), under the leadership of Japan. The principal regional cooperation structure in the Latin American region is the Space Conference of the Americas. Beginning in 2005, the African region started the African Leadership Conference on Space Science and Technology for Sustainable Development. The latter two regional conferences aim to raise awareness and the political profile of space among the governments of the region and have yet to establish operational space programmes. The Asia-Pacific entities have made somewhat more progress in this regard: APSCO is working towards a constellation of small satellites for environmental monitoring and disaster management, while APRSAF has implemented the Sentinel Asia programme for satellite-aided disaster management.

Fig. 2. *Second meeting of the interim Council of APSCO (source: CNSA).*

Fig. 3. *15th Session of APRSAF in December 2008 (source: JAXA).*

Another development is the growth in South–South collaborations among the intermediate and emerging space countries. An example of this is the China–Brazil CBERS cooperation in the development and operation of Earth observation satellites.

The established space-faring nations have very extensive cooperation programmes, and are also seeking to cooperate with intermediate and emerging space countries. Of course, cooperation is often driven by objectives such as strengthening strategic alliances, or using space to boost national prestige and regional influence, or to promote and sustain space industrial capabilities. All of these drivers lead in some cases to a sort of "competition to cooperate" with emerging space countries (Figures 2 and 3).

4.1.3.5. Rise of global utilities

The use of space technology is no longer the preserve of space experts. The development of intuitive, easy-to-use interfaces has seen enormous take-up of satellite broadcasting, communication and GNSS services by the general public. The take-up of Earth observation has been slower, but is developing at a rapid pace, as more intuitive interfaces (such as Google Earth) become available.

The easier access to space data and services has created greater demand for such data and services, thus fuelling the growth of profitable distribution channels. In the past, space agencies established entities to sell on a commercial basis data or services from agency-owned satellites. These operators soon realised that direct distribution space networks could generate large profits, so they started to acquire and operate their own space systems. Examples of this are satellite communications companies (such as Luxembourg-based SES Global) and more recently in the Earth observation domain the Google Earth partnership with GeoEye for acquiring its own data. The market for space data and services is now being shaped more and more by the entities operating the distribution channels to end users, rather than the providers of the space capabilities. Information distributors are now *owners* of space systems and are shaping user demand to create new markets for space-derived/delivered information.

The ready availability of ever-higher resolution imagery, satellite communication services and GNSS signals can promote as well as harm human security. From a space security perspective, military and security bodies will resist these developments, particularly in regard to imagery distribution, but the growing number of actors and the enormous reach of the internet will make it impossible to enforce restrictions indefinitely. Governments will have to face up to international public scrutiny and will no longer have a monopoly on the acquisition or interpretation of imagery. Already, there are a number of NGOs and news organisations that have made use of satellite imagery to document human rights abuses. This is surely an example of "fair and responsible" use of space.

4.1.3.6. Role of commercial space actors

The private sector is becoming an enabler of new space programmes. As of January 2008, 47 civilian entities had accessed space.[15] Quite a number of those did so with assistance of the commercial sector. The established Big Players in Europe and (to a lesser extent) North America are seeking emerging markets outside of their regions. Some are more constrained than others by regulatory barriers concerning technology transfer, such as the United States' International Traffic in Arms Regulations (ITAR). China and India are more recent entrants into the commercial space arena and can offer access to space or complete space systems delivered to orbit for emerging space countries. An example of this is the Nigerian communications satellite NigcomSat-1 built and launched for Nigeria by China's Great Wall Industry Corporation. In some cases, these actors are partnering with established large European players to offer "bundled" and/or "ITAR-free" services. In future, the personal spaceflight market will also provide space access opportunities to emerging nations.

Industry is starting to play a catalytic role in establishing international cooperation. An example of this is the Disaster Management Constellation, a cooperative project among Algeria, China, Nigeria, Turkey, and the UK in which Surrey Satellite Technologies Limited has played a catalytic and leading role.

If the history of human exploration and settlement is anything to go by, in future wealth will be created in space as well as on Earth. The commercial sector is today paving the way for these future developments, which will give rise to another set of debates on what constitutes "fair and responsible" uses of space. The history of the 1979 Moon Agreement, or, indeed, the 1982 United Nations Convention on the Law of the Sea, both with references to the Moon or the deep seabed and their resources as "the common heritage of mankind", shows that this will not be an easy debate.

4.1.3.7. Growing interest in space exploration in a country marks transition from emerging to intermediate space power status

Some emerging and intermediate space nations are beginning to adopt a view of space that goes beyond its purely utilitarian applications, providing an impetus for others to follow. There is a growing interest in manned spaceflight by countries that have not had this capability in the past. China has demonstrated human spaceflight capability and India is well on the way to doing so. This has given rise to other actors in the Asia-Pacific region initiating human spaceflight programmes, such as Korea and Malaysia, which have current astronaut development programmes taking advantage of flight opportunities offered by other nations. On the African continent, Nigeria has indicated an intention to do likewise in future. The burgeoning space tourism industry will provide more flight opportunities, not only for astronauts, but also for lofting small satellites or for performing scientific research in microgravity.

4.1.4. Preserving the space environment through "fair and responsible" use

4.1.4.1. Security on Earth underpinned by security in space

More and more, security on Earth is linked to security in space. As more States become actors in the space arena, the orbital environment will become a more crowded and complex environment in which to operate. To date, 29 States have demonstrated sub-orbital launch capability and 11 have demonstrated orbital launch capability. Security in space (just like security on Earth's roadways) will rely on the orderly and predictable behaviour of all users.

Perhaps the aspect of space security that is of most immediate concern to emerging space nations is that of preserving the Earth's orbital environment as a safe area in which to operate satellites, free from risk of disruption by space debris. This is an area in which the emerging space countries have a direct and critical contribution to make to "fair and responsible" use of space.

Space debris poses a serious threat to the space activities of developing countries, which may not be able to replace assets lost on orbit. The threat of impacts by debris will drive up development and insurance costs. The loss of an active spacecraft to debris impact may erode political support for space programmes in developing nations – operating satellites may be perceived to be too risky to justify the expected benefits of investment.

Because of the greater number of operational space systems in orbit, and the large and growing debris population, emerging space nations do not have the luxury of repeating the lessons learnt by the more established space nations in the early days of the Space Age. Emerging space nations should therefore take a number of steps to ensure that they are "fair and responsible" users of space. Such steps include:

- Developing space situational awareness capabilities, linked to those of other countries.
- Choosing launch service providers carefully and responsibly to minimise the chances of adding to the debris population.
- Adopting debris mitigation standards modelled on COPUOS/IADC voluntary guidelines, to be applied in the development and licensing of satellites, and in the selection of domestic and international component suppliers and launch service providers.

4.1.4.2. International cooperation as a means to preserve space security

One of the best means to enhance space security is to promote cooperation among the established and emerging space powers. Although bilateral cooperation is the favoured mode of cooperation among space agencies, large networks for collaboration (e.g., the Group on Earth Observation or the Global Exploration Strategy) lead to greater transparency and build confidence, mitigating mistrust and uncertainty as more countries gain access to space.

The established international legal regime for outer space activities (Table 1) provides another basis to promote transparency and responsible uses of outer space. The first four outer space treaties listed in the Table provide a basis for ensuring sustainable, equitable and secure access to space for current and future users of space. However, many States still have not acceded to these four treaties, including some COPUOS Member States. The fifth Treaty, the Moon Agreement, which

Tab. 1: *United Nations Outer Space Treaties.*

Treaty	Year	Ratifications
Outer Space Treaty	1967	98
Rescue Agreement	1968	91
Liability Convention	1972	87
Registration Convention	1975	48
Moon Agreement	1979	12

Source: United Nations Office for Outer Space Affairs.

attempted to deal with the notion of sharing the benefits of resource exploitation on the Moon, does not enjoy wide support among the leading space powers. Almost all of the ratifications of this Treaty are by countries not involved in lunar exploration. I will return to the issue of space exploration and developing countries in the last section of this paper.

The Outer Space Treaty, Rescue Convention and Liability Convention are all premised, to a certain extent, on the ability of States to identify the launching or responsible State for a given space object. The Registration convention makes provision for the identification of launching States with responsibilities for certain space objects. Out of 5734 payloads launched as of January 2008, 282 were not registered, and in recent years the trend of non-registration has been growing.[16] This should be a cause for concern to all space-faring nations.

While there is broad acceptance among States for the need to discuss "rules of the road" in the space arena, different States have different views concerning the implementation of such rules. Some States advocate non-binding, voluntary measures to address space security issues; others insist on a binding, treaty-based approach. Since the principal international space law-making body, COPUOS, operates on the basis of consensus, the effect of such disagreement could be to block progress on various fronts.

4.1.5. What about space exploration?

Through space exploration humanity is taking its first tentative steps from its cradle on Earth into the universe as a space-faring species. The immense public interest in international focus periods such as World Space Week in many developing countries with no space programmes demonstrates that people everywhere are excited by space exploration. So, what role is there, if any, for developing countries in the space exploration enterprise? In answering this question, we can recall that Article 1 of the Outer Space Treaty of 1967 reads as follows:

> "*The exploration and use of outer space, including the Moon and other celestial bodies, shall be carried out for the benefit and in the interests of all countries, irrespective of their degree of economic or scientific development, and shall be the province of all mankind.*"

It is interesting to note that space exploration is no longer the exclusive preserve of just a few major space powers. Some intermediate space powers and even some non-governmental entities (e.g., the Google Lunar X Prize and the International Lunar Observatory Association) are becoming active participants in the space exploration enterprise too. The appearance of these new actors potentially provides some new opportunities for emerging space nations to engage in the exploration enterprise.[17]

With more countries becoming involved in space exploration, it is clear that some form of coordination would benefit all stakeholders in the exploration enterprise. The Global Space Exploration Strategy[18] was adopted by the space agencies of 14 countries in May 2007. The Strategy is based on the premise of our common human destiny in space, as evinced in the opening remark of Chapter 1 of the strategy: "Space exploration is essential to humanity's future". In the closing remarks of Theme 4: A Global Partnership, the document states: "It [the strategy] is inclusive; the goal is to expand the opportunity for participation in space exploration to all nations and their citizens". However, the document does not address *how* to expand the opportunity for participation by developing nations, nor does it point to a process for doing so.

I would argue that, in its present form, the Global Space Exploration strategy is really an international strategy, not a global one. The word "global" is defined to mean "of or relating to the whole Earth". Yet, there were no agencies from Africa or Latin America listed among the organisations involved in drafting this document. Indeed, one would be hard pressed to find any emerging space nations represented among the countries that have developed this document. This lack of representation by a significant fraction of the world is reflected in the content of the document. In the 25-page document the phrase "developing world" occurs only once in a mention of a portable TB diagnostic tool developed as a spin-off of Mars exploration. The word "Africa" appears twice in the document in a reference to humans emerging from "ancient Africa".

So, is this an issue of "fair and responsible" use of space? I would argue that it is, because the "global" exploration enterprise seems to be leaving the developing world behind. For these countries, participation, even at a very modest level would have enormous national impact. Firstly, it would provide an opportunity for the scientific community in these countries to participate in a cutting-edge endeavour. Secondly, and perhaps more importantly, it would generally promote science and mathematics education in those countries, thus raising the general level of science and mathematics literacy. Some consideration should be given to creating mechanisms that would allow emerging space actors and even interested non-space-faring countries some opportunities to participate in the collective human adventure of space exploration, so that it becomes truly a global endeavour. Perhaps the main benefit of promoting a truly global space exploration agenda is that it would strengthen international cooperation in the peaceful use and exploration of outer space, thereby promoting a globally shared notion of the fair and responsible uses of outer space.

[11] United Nations. World Population Prospects: The 2004 Revision Highlight. United Nations Publication ESA/P/WP.193; Department of Economic and Social Affairs, Population Division. New York, United Nations, 2005.
[12] Peter, Nicolas. "The Use of Remote Sensing to Support the Application of Multilateral Environmental Agreements." Space Policy 20.3 (2004): 189–196.
[13] Peter, Nicolas. "The Changing Geopolitics of Space Activities." Space Policy 22.2 (2006): 100–109.
[14] Spacesecurity.org. Space Security 2008. Waterloo: Spacesecurity.org, 2008.
[15] Ibid.
[16] Ibid.
[17] Noumenia Project Team. Noumenia – Building on the Google Lunar X Prize: Recommendations for Future Activities on the Moon and Beyond. Strasbourg: International Space University, 2008.
[18] "Global Space Exploration Strategy: A Framework for Coordination." 31 May 2007. JAXA Press Release. 16 Mar. 2009. http://www.jaxa.jp/press/2007/05/20070531_ges_e.html.

4.2 Peaceful uses of outer space vs. militarisation: a cost–benefit analysis

Theresa Hitchens

4.2.1. Introduction

While questions regarding the future security of space often pit military uses against civilian – or "peaceful" – uses, reality is much more complicated. The world has benefited both from the militarisation of space and from progress in civil and commercial uses. Put simply, the use of space provides military (and thus security), economic and societal benefits.

Nor can the military and civil space spheres be easily separated. This is because space is a special environment, dominated by the laws of physics. Thus, both military and civilian assets in space must use similar orbits to accomplish different, but fundamentally similar, missions – such as Earth observation and communications. Further, due to the continued high costs of getting to orbit, many satellites, both those owned by commercial entities and those owned by governments, serve dual functions: providing services to both national militaries and civil society.

In addition, orbital assets face the same basic set of threats to their functionality and survivability from what is a fundamentally harsh environment. Space debris, solar flares, radio frequency interference all can create serious, or deadly, problems for satellites and spacecraft. For example, space debris – which is widely recognised as a growing problem because of the enormous damage it can wreak upon space craft upon impact – does not discriminate between military and civil satellites, or between satellites owned by one nation or another.

Finally, analyses of future space security often focus on the use of space by nation states – thus driving any debate down to the narrow concerns of national security. This often results in "zero sum game" thinking; i.e., the construct that one nation's gains in space capability, be it military or civil, results in a loss for another nation's security. But framing the debate in this manner misses a critical point: outer space is the ultimate globalised environment. Not only do about 50 nation states own and/or operate satellites, but also independent actors such as multinational firms, universities, academic consortia and even non-governmental organisations are increasingly active in either operating satellites or utilising data from them in novel (and non-state-centric) ways.

The bottom line is that actions by any one space sector, or even by any single space operator, affect all others. In particular, this raises questions regarding the trade-offs between actions that may improve military capabilities, but actually threaten civil and/or commercial capabilities.

4.2.1.1. Space is militarised but not weaponised

Another key factor in weighing how future uses of space will affect the security and sustainability of space operations is the fact that while the space environment can be said to be "militarised", it has not yet been "weaponised". That is, many nations and groups of nations use space for military purposes, but no nation has deployed dedicated weapons in space, or those designed to destroy satellites. This distinction is crucial. Arguably, the history of restraint among the world's space-faring powers regarding space warfare has allowed the rapid development of space activities that benefit humankind, in both the economic and scientific arenas. This is because the advent of anti-satellite (ASAT) and/or space-based weapons – and thus the subsequent threat of warfare in space – would dramatically increase the risks to spacecraft of all kinds – thus quite probably increasing the already high "costs of entry" for civil and commercial players.

Unfortunately, this situation may be changing due to improved technology, reductions in the cost of building and operating satellites and the increased perception of the military value of satellites among potential combatants. Evidence of this creeping trend towards weaponisation, after decades of relative quiescence in military space competition, includes: the Chinese testing of an ASAT weapon in January 2007; the U.S. decision to "shoot down" an ailing spy satellite in February 2008 using modified missile defence technology along with the strengthened U.S. doctrine of "space control";[19] and open debate in countries such as India, Israel and Russia about the potential value of ASATs and/or space-based weapon systems.

So, the central question to be addressed by this paper is more aptly: what are the costs vs. benefits for the sustainable and secure use of space by all stakeholders (military, civil and commercial) of increased militarisation and weaponisation of space? To answer this question, one first must review the benefits of space usage to the military, economic and civil society sectors; then consider the costs of increased militarisation and/or weaponisation to all three sectors.

4.2.2. Military benefits

It cannot be denied that outer space has been militarised since the dawn of the space age. During the Cold War, the Soviet Union and the United States actively, if often secretly, engaged in research and development of ASATs, space-based weapons and war fighting concepts that would utilise space as the proverbial "high ground" of battle. While the superpowers' interest in the potential of space warfare eventually ceded to concerns about the possible affect of space-related weapons on the arguably more important nuclear balance between the two sides, both Russia and the United States have continued to pursue separate military space programmes designed to enhance both strategic goals such as deterrence and performance on the battlefield. Meanwhile, many other nations have entered the fray: China, France, Germany, Italy, Israel, Spain and the United Kingdom all have, to a greater or lesser extent, dedicated military space assets. India, Japan, Iran and North Korea have similar potential, and some apparent interest in pursuing such a course.

Interestingly, in the recent past most military space programmes were dedicated to "strategic" purposes: missile and nuclear warning, intelligence gathering and verification of treaties and agreements (whether bilateral or multilateral) and communications. However, since the 1990s, military usage of space has shifted towards applications designed to enhance tactical operations on the battlefield. For example, the Global Positioning System (GPS) satellite navigation system is

A Space Enabled Reconnaissance-Strike Complex: The New American Way of War

KTO, 1991 (Desert Storm): 37 Days 1 Mbps/5K Forces	Unguided Laser/EO-guided	245,000 20,450	92% 8%
Serbia, 1999 (Allied Force) 78 Days; 24.5 Mbps/5K	Unguided Laser/EO-guided GPS-guided	16,000 7000 700	66% 31% 3%
Afghanistan, 2001-02 (Enduring Freedom) 90 Days; 68.2 Mbps/5K	Unguided Laser/EO-guided GPS-guided	9000 6000 7000	41% 27% 32%
Iraq, 2003 (Iraqi Freedom) 29 Days; 51.1 Mbps/5K	Unguided Guided	9251 19,948	32% 68%

Fig. 4. *Power Point Chart (courtesy of Lt. Col. (ret.) Peter Hays). Legend: KTO, Kuwaiti Theatre of Operations; EO, electro-optical; Mpbs, megabits; 5 K, bandwidth usage per 5000 troops.*[20]

increasingly used by the U.S. military to guide so-called "smart bombs" with greatly increased accuracy over older munitions using only inertial guidance systems, as well as for tracking U.S. forces (down to the individual soldier level) in the field (Figure 4).

Increased bandwidth for communications has led to a revolution in the amount of data that can be transmitted to commanders and forces in the field, including detailed imagery – the quality of which has improved greatly in the last two decades. Weather satellites further allow advanced planning in a way that would have been impossible in the Second World War.

In sum, space now provides huge tactical advantages to militaries who can harness it: improved battlefield awareness; 24/7 connectivity; rapid mobility; rapid and accurate strike capability; and fewer casualties, both to one's own military personnel as well within the civilian population.

There also are arguably new tactical advantages to be gained from the advent of ASATs and space-based weapons. ASAT weapons could eliminate some of the battlefield advantages provided by satellites illuminated above, as well as complicate long-range strike capabilities and make an attack more risky for the attacker. Further, some ASAT technologies – particularly ground-based kinetic energy (hit-to-kill) missiles – are already available and are relatively low cost compared to the price tag of space assets themselves. Space-based weapons (to target satellites, missiles or ground facilities) potentially provide global reach and 24/7 access to targets. They would further complicate the use of ground-based ASATs by an enemy, thus potentially reducing risks to one's own space assets. Therefore, it isn't difficult to understand why some military commanders are interested in such capabilities.

4.2.3. Economic benefits

It is obvious that the development of space as a commercial sector has brought economic benefits (as well as societal benefits, to be discussed below) to both those nations and companies operating satellites, as well as to the global public at large. What is actually harder to do is to quantify both the size of commercial space economy and specify the benefits resulting from that economic activity. This is largely because no nation state or global institution actually runs the traditional numbers on space commerce that are calculated annually for other economic sectors such as agriculture or the auto industry. Even the United States, which has the largest space economy (including military spending, spending by civil agencies such as NASA and the National Oceanic and Atmospheric Administration and commercial activity), keeps no statistics quantifying either the size of the military space budget, nor does it compile the basic facts about commercial space activities, services and benefits. According to a recent study by the non-governmental group Economists for Peace & Security: "The lack of reliable economic indicators represents an important gap in our knowledge abut the space economy and is a major impediment in the making of rational space policy."[21]

While some space-related industry groups and international institutions attempt to compile annual or semi-annual statistics, their methodologies differ as do their source materials. However, what can be seen from the statistics that exist is that space is a big business, and getting bigger every day. There are about 899 active satellites in orbit, with 390 owned by commercial firms or consortia and another 283 civil/non-military government-owned.[22] (That is, about two-thirds of the working satellites are not military owned/operated.) The Space Foundation, a U.S.-based non-profit dedicated to promoting the space industry, estimates that international revenue from government and commercial ventures at 251 billion U.S. dollars in 2007, up 11% from 2006.[23] The Organisation for Economic Cooperation and Development (OECD) estimated the global employment in 2006 in space industry manufacturing alone (with space services such as GPS receivers being the largest industry sector) at 120,000.[24] The U.S. Federal Aviation Administration (FAA), which is increasingly interested in commercial space activities, recently found that in 2006 commercial space transportation generated 139.3 billion U.S. dollars in economic activity and 729,000 U.S.-based jobs.[25]

There are also enormous indirect benefits to the global economy from the use of space, which are equally difficult to quantify. For example, the growth in global commerce spurred by the Internet as well as the emergence of the Internet-based economy. GPS – developed and operated by the U.S. Air Force – also has been crucial in enabling near-instantaneous financial transactions, improvements in the efficiency of air, ground and sea transportation, as well as better utilities management. The worldwide communications networks enabled by satellites have spurred globalisation of many industries and with that economic development, including in the poorest nations. Remote sensing – capabilities that also have their origins in military research – has enabled increased production of agricultural products. These economic benefits, in turn, lead to benefits to human society around the globe.

4.2.4. Civil society benefits

Examples of the benefits to civil society from the use of space are legion. They include the creation of jobs, in both developed and under-developed nations; weather prediction and disaster warning; disaster monitoring and response;

tele-education and tele-medicine for remote, poverty-stricken areas; communications in otherwise under-developed regions; climate change monitoring; refugee monitoring and assistance; documenting and monitoring wars and resulting crimes against civilians; resource management; improved scientific knowledge of

Angaba closeup before attack (source: eyesondarfur).

Angaba closeup after attack (source: eyesondarfur).

the Earth, the solar system and the universe; and increased international cooperation. As in the commercial sector, GPS is of particular importance for many civil society applications – as most computer operations rely on GPS clocks. Thus, it must be remembered that technological progress spearheaded by militaries often, and particularly in this case, spin off to other sectors of human activity.

Again, these benefits are next to impossible to quantify, but they are nonetheless tangible. For example, in recent years with the increased availability and thus reduction in costs of satellite imagery, non-governmental agencies such as Amnesty International have been tracking and documenting atrocities such as the ongoing genocide in Darfur.[26] The advent of tele-medicine has helped hundreds of thousands of patients from the poorest parts of India to the Australian Outback to rural Appalachia in the United States.

4.2.5. Costs to militaries of increasing militarisation/ weaponisation

While many nations have recognised the benefits that space assets can provide for military activities and capabilities, there are also costs inherent in embracing military architectures that rely heavily on space.

Most obvious is the fact that developing, building, launching and operating satellites remain expensive. Launch costs for the past two decades have hovered between 6000 and 17,000 euros per kilogram (depending on payload weight, size of rocket and the desired orbit). Indeed, according to commercial satellite operators, launch costs have actually gone up by about 50% over the past few years despite long-standing government and industry efforts to reduce them. Further, military satellites traditionally are large: 3–5+ metric tons each. Finally, satellite lifetimes are limited: 7–15 years – meaning recurring costs in replacing them.

Less obvious is the fact that reliance on satellites creates vulnerabilities for military operations. Satellites and spacecraft are fragile assets, vulnerable to damage from any number of threats – from space debris to ASATs. For example, NASA has had to replace at least 80 windshields on the Space Shuttle since the dawn of the programme due to impacts with debris.[27] Even the Hubble telescope has been damaged by debris, with one such collision resulting in a 1 cm hole in one of its gain antennas (Figure 5).[28]

They are also quite difficult to protect. Shielding against impacts by anything larger than 1 cm in diameter is quite simply impossible. Satellites travel is predictable orbits, and can be tracked – and if enough assets are dedicated to it, targeted (particularly those in low-Earth orbit which are relatively easy to reach

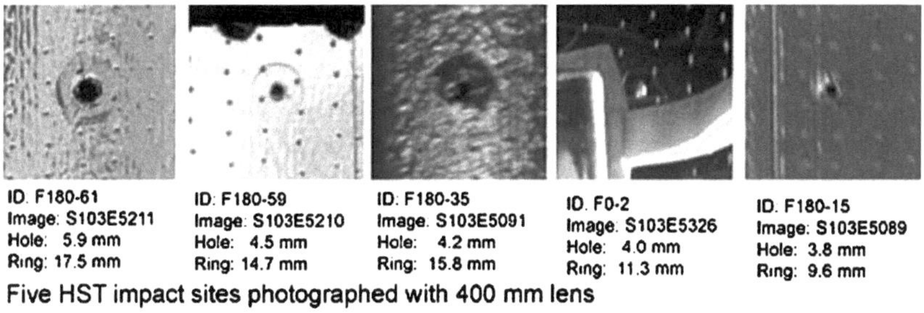

Fig. 5. *Holes in hubble telescope (source: NASA/JSC).*

using terrestrial-based missiles). Finally, the rapid diffusion of technology (ironically made possible by satellites!) means that advantages over potential adversaries are hard to maintain – thus ensuring that 'arms racing' in space is almost inevitable.

The spread of technologies applicable to ASATs has further increased the risks to military satellites. As satellites become more accessible as targets, the problem of reliance on them for critical military capabilities becomes more acute. The creation of dangerous debris from hit-to-kill ASATs – one of the most easily available methods for "killing" a satellite – adds to the threat picture. Finally, the potential use of ASATs in a conflict raises the likelihood of rapid conflict escalation, perhaps even up to the use of nuclear weapons.

And while space-based weapons might provide some advantages to achieving particular military missions, they would be expensive – and for many military missions the cost/benefit ratio wise other means of attack is negligible if not actually negative. Of course, space-based weapons are also just as vulnerable as other satellites. Perhaps most insidiously, the deployment by one nation of space-based weapons would increase incentives for others to develop means to target space-based objects, thus increasing the risks to all objects on orbit (including non-military) since it is fiendishly difficult to discriminate a space-based weapon from a perfectly benign satellite.

4.2.6. Commercial costs of increased militarisation/ weaponisation

The cost for development and production of a typical commercial communication satellite (according to operators) is between 140 million U.S. dollars and 180 million U.S. dollars (108 million to 139 million euros), and the launch adds another 120 million U.S. dollars to 140 million U.S. dollars (93 million to 108 million euros). Obviously, the loss of any one satellite due to acts of war would be a loss of this investment plus the loss of planned business revenue.

Unfortunately, the actual impacts of increased risks caused by further militarisation or weaponisation of space could be felt by commercial and civil operators even before any "shots were fired" in space.

A first, and well documented, impact of increased militarisation of space is restrictions on trade in space-related technologies that can hamper the ability of commercial companies to do business in an extremely globalised market. This dilemma is most obvious in the United States, where the subjection of commercial space technology exports to the International Traffic in Arms Regulations (ITAR) severely limits the transfer of technology and know-how. Under ITAR rules, if a non-U.S. firm buys any U.S.-made space-related systems or even parts, its future exports also become subject to U.S. export control. This is particularly problematic for trade with China, as China is the primary subject of U.S. export control concern in space – and the reason ITAR rules were slapped on space technologies in 1999.[29]

Space industry representatives and supporters of new space ventures such as tourism – from the American Institute of Aeronautics and Astronautics (AIAA) to the Space Foundation – have all slammed the ITAR regime as unworkable, and as reducing U.S. market share dramatically over the past decade.[30] Charles Huettner, executive director of the Aerospace States Association, said in July 2008, "ITAR has led to increased global competition and is a significant impediment to the U.S. space industry's ability to market to foreign buyers resulting in decreased sales and competitiveness".[31] Of increasing worry to U.S. satellite manufacturers is the successful move by European companies EADS and Thales Alenia Space to market "ITAR-free" satellites and large subsystems such as motors, even if they are more expensive than those made with U.S. components – with some efforts to develop European-only critical space components even being funded by the European Space Agency (ESA).[32]

Not so obvious is the pressure increased militarisation or weaponisation may place on operators who must weigh enormous investment costs against the likelihood of loss. In particular, insurance for launch and initial on-orbit operations increases this price tag by 8–10% currently, and that percent would of course go up in an environment considered more risky by insurance companies.

Most worrisome to commercial operators is the spectre of space warfare where not only are their own satellites targets (as many commercial satellites provide services to militaries and thus might be considered fair game in a fire-fight), but also where the combatants are using kinetic energy ASATs that create enormous amounts of debris. Even the destruction of a handful of large military satellites in low-Earth orbit could result in increases to the debris population that could render entire orbital bands unusable, or impassable, by spacecraft (Figure 6).[33]

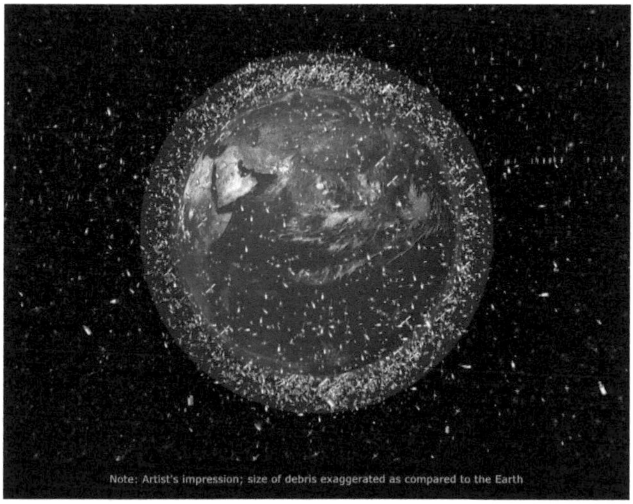

Fig. 6. *Current debris objects in LEO (source: courtesy, ESA).*

According to space market analyst Marco Cacéres of Teal Group;

> *"About the last thing that the satellite market needs now is the uncertainty that will accompany any moves to start blowing up objects in space or arming military satellites with protective countermeasures. The added debris problem is bad enough. An ASAT weapons race will have the effect of increasing the financial risk of any satellite programme, and undoubtedly be felt most within the commercial market through decreased investor confidence and (or) higher insurance rates".*[34]

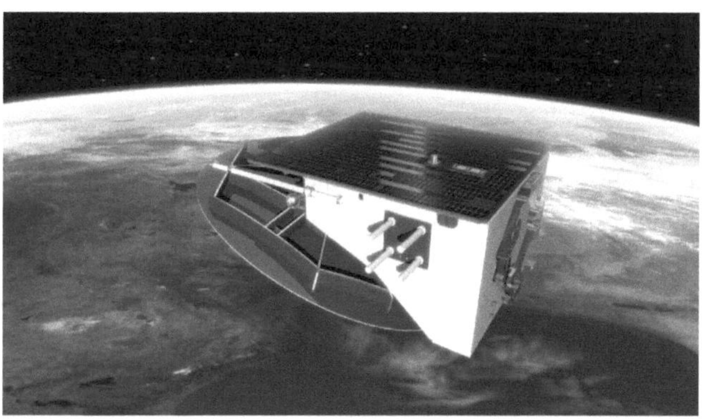

Fig. 7. *SAR-Lupe Illustration (source: courtesy, OHB System).*

Finally, another factor that comes into play is the willingness of militaries to utilise the services of commercial providers if the space environment is considered a militarily contested arena. More and more nations seem to be moving towards dedicated military satellites, with the most recent example Germany's SAR-Lupe radar satellite constellation commissioned by the German military on 4 December 2008 (Figure 7).[35]

In another example, the U.S. Defence Department decided in September 2008 to buy two new imagery satellites, dubbed BASIC for Broad Area Space-Based Imagery Collection, with capabilities no better than that provided by commercial companies GeoEye and Digital Globe, due to concerns about "ownership" and "control" of the satellites and image gathering.[36] That plan, however, was shot down by Congress due to concern that it violated presidential policy regarding government use of commercial capabilities where feasible as well as the likely price tag.[37] Considering that the U.S. military since 2005 has been the main consumer of commercial satellite communications, accounting for about 90% of overall revenues,[38] and more than half the market for commercial imagery comes from government contracts,[39] movement by militaries to procure more dedicated capabilities could affect the commercial market in a negative manner.

4.2.7. Costs to civil society of increased militarisation/weaponisation

Perhaps the least quantifiable, but most serious, cost to civil society should the trend towards greater militarisation and weaponisation of space continue would be the increase in tensions among the world's space powers regarding space capabilities. Fears that other nations might gain military advantage would likely dampen political will for international cooperation – which is becoming increasingly vital in areas such as disaster and climate monitoring. Cooperation on space exploration would also be affected.

Indeed, the U.S. ITAR rules already have undercut efforts by NASA to forge strong space science and exploration partnerships, and engendered considerable hostility among European civil space agencies. A key example of how ITAR can actually sabotage civil collaboration was NASA's failed 2005 Demonstration of Autonomous Rendezvous Technology (DART) mission, involving a small satellite designed to closely orbit another satellite. NASA's "mishap report" on the mission highlighted the fact that "perceived restrictions" under ITAR resulted in "insufficient technical communication between the project and the international vendor" in part led to a lack of understanding of the spacecraft's parameters and the mission risks.[40]

ITAR also was blamed for complicating NASA's collaboration with Europe on the Automated Transfer Vehicle for the International Space Station, causing NASA Administrator Michael Griffin to write to U.S. Secretary of State Condoleezza Rice in April 2007 seeking relief from the regulations.[41] Indeed, some European scientists were prohibited from using the computers in NASA's Mars Exploration Rover science operation facility, even though they were supposed to be developing computer command and control software for the instruments ESA had built and provided to the mission.[42]

The situation could only get worse if military tensions are heightened and concerns about protecting national technical advantages rise. Further, deployment of space weapons would almost inevitably exacerbate those negative trends.[43]

The balance of government investment in military wise civil space programmes is also a potential problem if nation states become more focused on military uses of space assets. While relatively few nations currently maintain separate military and civil space programmes (chiefly, the United States, Russia and France), the trend towards integrating space capabilities with military forces – a trend that is spreading far and wide even to traditionally pacifistic nations such as Japan – inevitably will lead to internal battles for resources. For example, in the United States, NASA's budget has remained essentially stagnant over the last decade, while known military space spending has been growing especially in the past few years. While determining actual national security space spending in the United States is nearly impossible due to vague reporting, the multiplicity of agencies and military services with space-related budgets, and classification. But by at least one estimate, national security space spending jumped from about 20 billion U.S. dollars in fiscal year 2005 to about 30 billion U.S. dollars in fiscal year 2008.[44]

As with commercial vendors, civil government agencies also have to weigh the potential risks with the costs and benefits of developing and deploying on-orbit assets. A riskier environment caused by increased tensions, and the advent of space-related weaponry, would certainly raise the bar for agencies (and parliaments) weighing public investments in satellites and spacecraft.

In a similar vein, the advent of hostilities in space as part of traditional warfare would place civil space assets and services at dire risk – especially if debris-creating weapons were involved. This is not a trivial concern, given the reliance of modern societies on space-enabled capabilities.

4.2.8. Conclusions

Although much more detailed study would be required to complete a detailed and quantified cost–benefit analysis of increased militarisation and weaponisation of space, several conclusions can be drawn from even this cursory overview.

First, growing interest of militaries around the world in capturing the benefits of space-enabled operations – of which there are many – threatens to raise tensions among space-faring powers.

Second, increased military tensions in space will lead (and in some cases already is leading) capable nations into consideration of the potential advantages of ASATs and space-based weaponry, and the potential development and deployment of such weapons.

Third, trade-offs (in budgets and risk levels) between military advantages and commercial/civil disadvantages will be required if these first two trends continue. That is, commercial and civil space efforts almost inevitably will suffer from increased investments in military space capabilities and the subsequent increase in tensions and risks.

Fourth, space weaponisation and/or space warfare – especially if destructive weapons are involved – would increase the risks to all space assets and operations, whether military, civil or commercial.

Finally, and most importantly, it is almost impossible not to conclude that the costs of space warfare would exceed any benefits – since any benefits would be only short-term tactical military benefits whereas costs would be long-term and affect all sectors of human space activity.

Thus, it behoves space powers to move cautiously in the military space sphere and to undertake efforts to develop holistic national space strategies that take into account the interlocking nature of military, civil and commercial space activities. Balance and prudence will be necessary watchwords if the space environment is to remain sustainable and secure for use by future generations.

---

[19] The 2006 U.S. National Space Strategy details a historically robust concept of 'space control' that stakes out U.S. rights to 'freedom of action' in space as well as claims a U.S. right to: "dissuade or deter others from either impeding those rights or developing capabilities intended to do so; take those actions necessary to protect its space capabilities; respond to interference; and deny, if necessary, adversaries the use of space capabilities hostile to U.S. national interests". See: White House Office of Science and Technology Policy "U.S. National Space Policy". 2006. 4 Nov. 2009. http://www.ostp.gov/galleries/default-file/Unclassified%20National%20Space%20Policy%20–%20FINAL.pdf.

[20] Sources for Figure 1 include: Central Air Forces. "Operation Iraqi Freedom: By the Numbers." Presentation. Prince Sultan Air Base, Saudi Arabia. 30 April 2003; Lambeth, Benjamin S. The Transformation of American Airpower. Ithaca: Cornell University Press, 2000; and Cohen, Eliot A. Gulf War Air Power Survey, Summary Report. Washington, D.C.: Government Printing Office, 1993.

[21] Kaufman, Richard, Hertzfeld, Henry, and Jeffrey Lewis. Space, Security and the Economy. Washington, D.C.: Economists for Peace & Security, September 2008: 4.

[22] Union of Concerned Scientists. UCS Satellite Database. 6 June 2008. http://www.ucsusa.org/nuclear_weapons_and_global_security/space_weapons/technical_issues/ucs-satellite-database.html.
[23] Space Foundation. The Space Report 2009. Colorado Springs, CO: Space Foundation, 2009: 6.
[24] Kaufman, Richard, Hertzfeld, Henry, and Jeffrey Lewis. Space, Security and the Economy. Washington, D.C.: Economists for Peace & Security, September 2008: 8.
[25] Federal Aviation Administration. The Economic Impact of Commercial Space Transportation on the US Economy. Washington, D.C.: FAA. April 2008: 1.
[26] Amnesty International website. Eyes on Darfur. 4 Nov. 2009. http://www.eyesondarfur.org/satellite.html.
[27] Atkinson, Nancy. "Space Debris Illustrated: The Problem in Pictures." 11 April 2008. Universe Today. 4 Nov. 2009. http://www.universetoday.com/2008/04/11/space-debris-illustrated-the-problem-in-pictures.
[28] Lepisto, Christine. "Shocking Space Debris Images." 20 April 2008. Treehugger. 4 Nov. 2009. http://www.treehugger.com/files/2008/04/shocking-space-debris-images.php.
[29] Secure World Foundation website. International Traffic in Arms Regulations (ITAR) Factsheet. Last updated September 2008. 4 Nov. 2009. http://www.secureworldfoundation.org/siteadmin/images/files/file_229.pdfm.
[30] See: "ITAR and the U.S. Space Industry." September 2008. Space Foundation. 4 Nov. 2009. http://www.spacefoundation.org/docs/SpaceFoundation_ITAR.pdf; and Messier, Doug. "AIAA Supports Loosening of U.S. ITAR Export Regulations." 11 Mar. 2008. Parabolic Arc. 4 Nov. 2009. http://www.parabolicarc.com/2008/03/11/aiaa-supports-loosening-of-us-itar-export-regulations/.
[31] Kuslolek, Richard. "ITAR Dilemma: Finding the Balance Between Regulation And Profit" 1 July 2008. Satellite Today. http://www.satellitetoday.com/military/homelandsecurity/23649.html.
[32] Brown, Peter J. "Engaging China in Space Part 2: US Firms Tired of Being Shut Out." 4 Dec. 2008. Asia Times. 4 Nov. 2009. http://www.atimes.com/atimes/China/JL04Ad01.html.
[33] Wright, David. "Space Debris." Physics Today 60.10: 35–40.
[34] Cáceres, Marco. "Market Impact Brief: China's ASAT Weapons Test." January 2007. Teal Group Corp. Arlington, VA.
[35] "Germany Commissions First Spy Satellite", 4 Dec. 2008. Deutsche Welle DW-World. 4 Nov. 2009. http://www.dw-world.de/dw/article/0,,3849862,00.html.
[36] Shalal-Esa, Andrea. "U.S. Approves Plan for 2 Commercial Image Satellites." 10 Sept. 2008. Reuters. 4 Nov. 2009. http://www.reuters.com/article/reutersEdge/idUSN1045748920080910.
[37] Gardener, David W. "Congress Cuts BASIC Satellite Imaging Program." 21 Oct. 2008. Information Week. http://www.informationweek.com/news/showArticle.jhtml?articleID=211300121.
[38] "Market to Generate $7.8 Billion in Satellite Communications Revenue Through 2010." 15 Mar. 2005. Northern Sky Research Press Release. 4 Nov. 2009. http://www.spaceref.com/news/viewpr.html?pid=16401.
[39] Eisler, Peter. "Global Security: The Sky's Unblinking Eyes." 9 Nov.2008. USA Today. 4 Nov. 2009. http://www.technewsworld.com/story/65090.html?wlc=1229102384.
[40] NASA. NESC Review of Demonstration of Autonomous Rendezvous Technology (DART) Mission Mishap Investigation Board Review (MIB). NASA Engineering and Safety Center Report: Demonstration of Autonomous Rendezvous Technology Mishap Review Board. Doc. No. RP-06-119, Version 1.0. December 2006: 31.
[41] Berger, Brian and Peter B. de Selding. "ITAR Complicates Preparations for New Station Supply Vehicle." Space News 11 Apr. 2007.
[42] Lakdawalla, Emily. "A Backlog of 400 Emails: ESA and NASA; MGS Failure; Iapetus; Saturn's D Ring." 25 July 2007. The Planetary Society Blog, The Planetary Society. 4 Nov. 2009. http://www.planetary.org/blog/article/00001044/.
[43] Assembly of the Western European Union. Weapons in Space. Fifty-Second Session. Document A/1932 of 21 June 2006. Paris: WEU: 7.
[44] Kaufman, Richard, Hertzfeld, Henry, and Jeffrey Lewis. Space, Security and the Economy. Washington, D.C.: Economists for Peace & Security, September 2008: 10–16.

4.3 The space debris environment and its impacts

Fernand Alby

4.3.1. The situation in orbit

The problem of space debris first arose on 4 October 1957. On that date, the Soviet Union placed the very first artificial satellite in orbit around the Earth using a Semyorka rocket. This rocket's final stage (6500 kg) and protective fairing (100 kg) remained in the same orbit as Sputnik (84 kg), so in fact the 'functional' payload was little over a mere 1% of the total mass injected into orbit. Moreover, this 1% operated for just 21 days before re-entering the atmosphere 92 days later: the "functional" 1% of the injected mass had thus been debris for three quarters of its orbital lifetime.

Since then, space activities have developed considerably and the population of objects in orbit has not stopped growing: human activity has led to the proliferation in space of a huge number of objects of all sizes. Recent calculations estimate around 12,000 objects measuring over 10 cm in size, 200,000 objects between 1 and 10 cm and 35,000,000 objects between 0.1 and 1 cm. Particles measuring less than 0.1 cm are even more abundant. For almost any size of object in space, man-made pollution now represents a greater risk than the meteors found in the "natural" space environment.

These objects derive from a number of sources (see Figure 8):

- Operational satellites, which number around 600, and satellites at the end of their lifetime that remain in orbit around the Earth.
- Upper stages of launchers that have been used to place these satellites in orbit.
- Operational debris, objects intentionally released during a mission: casings needed to protect instruments during the launch phase, mounting systems for solar panels or antennas before their deployment in orbit, release mechanisms, straps, etc.
- Fragmentation debris: debris produced after a collision between an object in orbit and space debris or meteorites. Also, debris resulting from spacecraft accidentally or intentionally exploding.

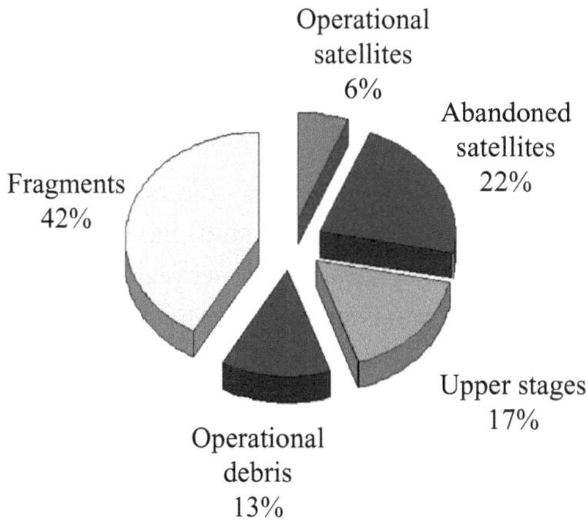

Fig. 8. *Debris' categories: main object categories in orbit (source: CNES).*

- Propellant residues: solid propellant motors that are used to carry out orbit transfers, particularly between a transfer orbit and geostationary orbit, release small alumina particles during thrust. This problem is especially critical at the end of the thrust when combustion becomes unstable and slag measuring several centimetres may be ejected into space.
- Ageing of materials in space. The space environment is extremely hostile to acute temperature changes in areas exposed to shade then sunlight, atomic oxygen, ultraviolet rays, etc. This ageing leads to large quantities of debris being produced (photoelectric cells becoming detached, heat shield covers flaking, paintwork peeling off, etc.).

There are other, more anecdotal sources of debris, which can also have a significant impact on the population of objects in orbit:

- In 1961 and 1963 the U.S. Air Force planned to release several million copper needles (Westford Needles) into orbit at an altitude of around 3000 km, as part of the Midas 4 and Midas 6 experiments. The objective was to create a ring of dipoles around the Earth that could be used as a passive reflector for military communications. Only the second experiment was partially successful. These needles then formed into clusters: 65 of these clusters could be observed from the ground even as late as 1998.
- In the 1980s the Soviet Union operated nuclear-powered RORSAT ocean reconnaissance satellites. At the end of their mission, their nuclear reactor cores were re-orbited at altitudes between 900 and 1000 km to allow their radioactivity to decrease before they fell back into the atmosphere. Leaks in the cooling

circuits were observed on 16 of these satellites, which resulted in thousands of droplets of liquid sodium and potassium being released into orbit. These droplets measure between 1 mm and a few centimetres.

Since 1957 the growth in the number of objects catalogued over time (those objects that can be monitored from the ground) has been more or less constant: they are increasing in number by around 220 per year. Future growth will depend essentially on the number of launches and the number of objects placed into orbit by each launch: over the last few years the number of launches has dropped, mainly due to a decline in activity in Russia, although other factors have had the opposite effect. Plans for constellations involving several dozen, or even several hundred satellites, have raised fears of a rapid increase in population. These projects are currently on hold but another potential risk is beginning to emerge with the development of "small" satellites (micro, nano and pico satellites) that could be launched in clusters by a single launcher. Another significant factor is the effectiveness of mitigation measures, which are being applied with increasing frequency. Anti-satellite tests such as the destruction of Fengyun 1 C in January 2007 and USA 193 in February 2008 have, of course, a negative impact on the environment. The consequences are particularly important when these tests are conducted at a relative high altitude (case of Fengyun 1 C at 850 km) that leads to a large number of debris having a very long orbital lifetime.

With large enough objects (those typically measuring 10 cm in low orbit and between 0.5 m and 1 m in geostationary orbit), the name, origin, and orbit parameters of each one are catalogued, enabling their trajectories to be predicted. These objects are regularly observed by radars in low orbit and telescopes in higher or geostationary orbits. The American SSN (Space Surveillance Network) has thus been able to compile a catalogue of around 12,000 objects. Smaller objects can also be observed, but as their trajectory cannot be calculated, it is impossible to find them again later. Their observation only provides statistical information on the number and size of objects to be obtained. Lastly, very small objects (dust) cannot be detected using ground-based facilities. Satellite-borne detectors are used to measure flows of small particles. Post-flight analysis of control surfaces exposed to the space environment then brought back to Earth after a mission have also provided a wealth of information on the number, artificial or natural origin, and mass of particles encountered. This was the case, for instance, with the solar panels from the Hubble telescope that were recovered in orbit on several occasions, and with various surfaces fitted to the space stations (International Space Station (ISS) or MIR) and to dedicated vessels such as the Long Duration Exposure Facility (LDEF). It was also the case with the Space Shuttle, which undergoes a detailed inspection after every flight.

The distribution of debris in space is not uniform, but obviously its concentration is greater in "useful" orbits where human activity is greatest, particularly in the geostationary orbit where most of the telecommunications satellites are found and in low orbits between 600 and 1500 km that correspond to many Earth observation missions.

The orbital lifetime of these objects is limited by the influence of the atmosphere. Atmospheric density diminishes more or less exponentially as altitude increases and the residual atmosphere found in low orbits has a decelerating effect on objects in orbit. The consequence of this deceleration is to lower an object's altitude and therefore increase atmospheric friction, which eventually results in objects falling back down to Earth. For example, at the altitude of the ISS (around 350 km), an object's lifespan is just a few months, which requires manoeuvres to be carried out regularly to compensate for this effect. However, this phenomenon only affects objects in low orbits: even at SPOT's altitude (800 km) the lifespan is of the order of one or two centuries. In higher orbits, lifespan can be measured in millennia or tens of millennia. With elliptical orbits, such as the transfer orbits which are used to reach geostationary orbit, an object's lifespan depends essentially on its perigee altitude: for a perigee altitude of 200 km the orbital lifetime ranges from a few months to a few years. A perigee altitude above 600 km results in orbital lifetimes measuring in the millennia. Lastly, in geostationary orbit, where there is no trace of any atmosphere, lifespan has no limits on a human scale.

4.3.2. Risks in orbit and on the ground

Obviously, this debris does pose a risk in the event of a collision with operational satellites. In orbit, these objects move at relative speeds of as much as 15–20 km/s. At these rates, even small particles have considerable kinetic energy: there is currently no shielding that can resist objects measuring more than 1 or 2 cm.

Impacts caused by small pieces of debris can be seen on any surface that has spent time in space and been brought back to Earth. For example, impacts were noted on the Shuttle Endeavour after flight STS 118 in August 2007: a perforation measuring 8×6 mm was observed on a radiator panel. The astronauts themselves have been able to observe damage to the ISS during their spacewalks (e.g., a torn thermal cover was discovered on the Zarya module in June 2007). With the Space Shuttle, the post-flight inspection leads to an average of one window per mission being replaced due to impact damage.

These data have made it possible to calculate that, statistically, the ISS may be struck by an object measuring over 1 cm every 71 years and that the Hubble telescope, during its theoretical 17-year lifespan, has a 4% chance of experiencing

Tab. 2: *Average time between two debris impacts in low orbit on a satellite measuring 100 m² (source: ESA).*

Altitude (km)	Debris measuring 0.1 mm (days)	Debris measuring 1 mm (years)	Debris measuring 1 cm (years)	Debris measuring 10 cm (years)
400	10	3	885	12,900
780	1.5	1	155	1190
1500	1.6	1.6	270	1590

the same kind of impact. Table 2 gives the average time between two debris impacts on a 100 m^2 satellite according to its altitude and the size of the particles (see Table 2).

The consequences of these collisions depend on the impact site: on a satellite, perforation of a solar panel, an antenna or even a wall is generally of no importance. However, a high-speed impact between a small object and a fuel tank or an electronic unit could result in the satellite being lost.

The first official collision in space between catalogued objects took place on 24 July 1996, when debris from the in-orbit explosion of the third stage of Ariane flight V16 severed the stabilisation mast on the Cerise microsatellite (Figure 9). Two other collisions between catalogued objects were subsequently revealed: a collision on 17 January 2005 between a stage of the American Thor launcher and a fragment from a Chinese CZ-4 launcher stage and a collision between the Russian Cosmos 1934 navigation satellite and debris from the Cosmos 926 satellite, which occurred in December 1991 but was only identified in 2005. Unfortunately, the

Fig. 9. *"Debris' collision Artist's view" (source: CNES).*

likelihood of this kind of event occurring will increase in the future due to the growth in the population of objects orbiting the Earth.

At the moment, objects measuring between 1 and 10 cm represent the greatest danger as shielding is unable to stop them and it is impossible for satellites to avoid them because they are too small to be tracked from the ground.

Space debris is also a potential hazard on the ground: objects in low orbit are slowed down by the residual atmosphere and eventually fall back to Earth. Most of these materials disintegrate during re-entry because of the extremely high temperatures but some elements can survive these conditions and reach the ground. For example, with the MIR space station, whose mass in orbit was 140 tons, Russian specialists estimated that 20% of its mass would survive as debris, representing an excessive risk. The decision was therefore taken to conduct a controlled re-entry to ensure that the debris fell over the South Pacific. There are two kinds of re-entry. With uncontrolled re-entries, an object falls anywhere within the latitudes corresponding to its orbit inclination. When the risk to inhabitants is too great, a controlled re-entry must be conducted. One or more manoeuvres are needed to ensure that the object falls in a precise place (i.e. the ocean) in order to minimise the risks. In general, agencies consider the level of acceptable risk to be around 10^{-4}, or a few 10^{-4} (probability of there being a victim during the operation). When the risk is below this threshold, a natural uncontrolled re-entry is acceptable. However, if the risk should exceed this threshold, controlled de-orbiting is essential to bring the level of risk back down.

Currently, one to two catalogued objects fall back down to Earth each week and pieces from these objects are regularly recovered from the ground: some of these (helium tanks, fuel tanks, engine combustion chambers, etc.) may be masses weighing several dozen kilograms. To date, no casualties have been reported as a result of falling space debris.

4.3.3. The solutions

4.3.3.1. Evaluation of the available solutions

To deal with a situation which is of growing concern, there are four potential solutions: clean up space to reduce the amount of debris, use shielding to protect objects from impacts, avoid debris and reduce production of debris (prevention). We shall see that the first three solutions do not work, or work only partly, and therefore the only answer is prevention.

First, the solutions that do not work. There is no way to clean up orbits by eliminating debris. Because of the high speeds at which orbiting objects move (several km/s), any capture system such as the "butterfly net", absorbent foam, etc., would simply result in the intercepting system and the debris disintegrating upon collision. Ground-based destruction systems based on powerful lasers have also been considered. Apart from the fact that their feasibility is far from proven (power needed, pointing accuracy, uncertainty about the trajectory of the debris), the experts agree that it would be better to have one intact object in orbit that can be tracked from the ground than hundreds or even thousands of smaller pieces of potentially dangerous debris.

Another alternative would be to recover debris in orbit using a vessel such as the Space Shuttle. This kind of solution would require a certain number of technical problems to be overcome, for example how to perform a rendezvous with an uncooperative object, probably rotating, with an uncontrolled attitude; then how to grasp it and secure this potentially dangerous object (perhaps containing residual fuel and thus representing an explosion hazard) in the cargo bay; and then how to carry out atmospheric re-entry with such a cargo. Besides, the Shuttle's capabilities are limited to altitudes below around 600 km and slightly inclined orbits. Furthermore, after completing its first rendezvous, it would be out of the question, due to the available fuel, to modify its trajectory plan and seek another object located in a different orbit. Under these conditions, the cost of such a mission to retrieve a single piece of debris, or a few pieces of debris located in adjacent orbits, would seem exorbitant.

Some partial solutions are available. It is possible to protect spacecraft using shielding. However, given the speeds in orbit and the corresponding energy, no shield is able to stop particles measuring more than 1 or 2 cm. Moreover, shields add significantly to a spacecraft's mass, so their use is currently reserved for permanently-crewed space stations.

Another partial solution consists in avoiding collisions when the debris' trajectory is well understood: avoidance is theoretically feasible whenever there is a risk of collision with catalogued debris. The process is still difficult, however, because the catalogues are not sufficiently accurate and radar facilities must also be used to reduce uncertainty and limit false alarms. Furthermore, it has to be possible to detect the collision several days in advance to have time to carry out the necessary analyses, conduct measurements, confirm the risk and take any decision. It also generally requires the mission to be interrupted until the satellite has returned to its initial orbit. This process is indispensable for manned vessels such as the ISS or the Space Shuttle. For example, on the ISS such close surveillance led to the following eight avoidance manoeuvres being conducted:

- 27-Oct.-1999 ISS-Pegasus Rocket Body
- 30-Sept.-2000 ISS-Vostok Rocket Body

- 10-Feb.-2001 ISS-Space Shuttle Elektron 1 Debris
- 14-Mar.-2001 ISS-Space Shuttle ISS/Shuttle Debris
- 15-Dec.-2001 ISS-Kosmos Rocket Body
- 16-May-2002 ISS-Kosmos Rocket Body
- 30-May-2003 ISS-Megsat
- 28-Aug.-2008 ISS-Cosmos 2421 debris

It should also be noted that surveillance potentially leading to avoidance manoeuvres is being implemented more and more with reference to satellite control (e.g., CNES currently has 15 satellites under surveillance).

The only solution that can be applied immediately, therefore, is prevention: this means no longer creating any space debris, or creating as little as possible. These measures aim to reduce or stabilise the rate of population growth of objects in orbit.

4.3.3.2. Prevention measures

Priority is given to applying these measures to the two most crowded and hence most polluted zones in space (see Figure 10):

- The low orbit zone: altitudes below 2000 km,
- The geostationary zone: a corridor extending ±200 km each side of geostationary altitude and limited to ±15° of inclination.

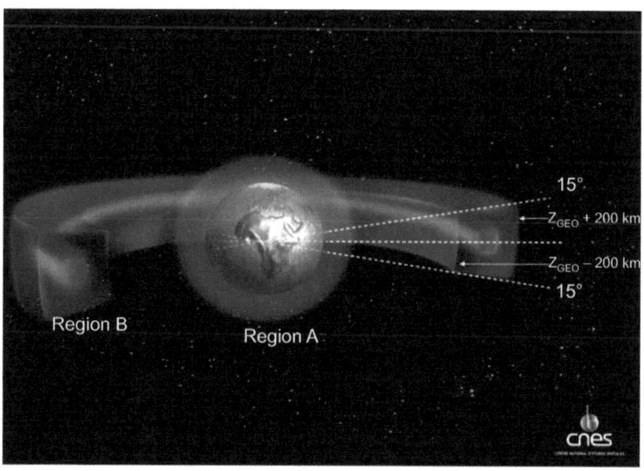

Fig. 10. *Zones where the measures will take place: protected regions in space (source: CNES).*

4.3.3.2.1. Principles of the measures

The prevention measures can be summarised according to four major principles:

- Do not intentionally release objects into space: covers, hoods, straps, springs and fastening mechanisms used during launch must all be "trapped", i.e. they must remain fixed to the satellite and must no longer be abandoned in orbit as was often the case before. Pyrotechnic cutting must be performed cleanly and the debris created by rupturing materials must be captured. Solid propellant motors used for positioning operations can also be a significant source of debris: some models release particles throughout combustion, particularly at the end of combustion with the low-speed ejection of particles measuring around a centimetre, which remain in orbit.
- Reduce the probability of explosions in orbit: the fragmentation of a spacecraft in orbit leads to a huge amount of debris of all sizes being created and in-orbit fragmentation thus represents the main source of debris. This constraint is generally taken into account by the designer and the operator during the mission to ensure the necessary reliability. However, once the mission has been completed, this constraint disappears and the operator must take special measures to avoid any subsequent explosion of the satellite or launcher that it will be abandoning in orbit. These operations, known as passivation, consist in reducing all sources of on-board power: by emptying fuel tanks, opening pressurised gas tanks, discharging batteries, etc.
- End-of-life manoeuvres: the aim of these measures is to remove the decommissioned vessel from protected regions. In low orbit the generally accepted requirement is the 25-year rule: no vessel may remain more than 25 years in the protected region after its operational mission has ended. This can be accomplished in a number of ways including: direct (controlled) de-orbiting, indirect de-orbiting (lowering the perigee so that atmospheric friction causes the object to fall back to Earth in less than 25 years), moving the vessel up above the protected region, or using another spacecraft to recover it. In geostationary orbit, this solution is impractical due to the region's remoteness from Earth. The recommended solution is thus to transfer the vessel into a "graveyard" orbit located 200 km above geostationary orbit, where it can no longer obstruct the protected region. After the de-orbiting or re-orbiting manoeuvres have been completed, the vessel must be passivated to avoid any subsequent risk of explosion.
- Prevention of in-orbit collisions: a collision between two vessels would generate a massive amount of debris. This risk can be decreased by the choice of mission orbit (choosing a relatively unpopulated orbit), the choice of launch time with regard to launch-related risks, and by conducting in-

orbit avoidance manoeuvres. Employing the latter two measures is only possible if the operator has enough information to calculate the collision risks, which is generally not the case: the catalogue of objects in space published by the United States only contains data on some of the objects over a certain size and the precision of these data is somewhat inadequate. The operator therefore needs access to other facilities (i.e. military) in order to accomplish this function.

4.3.3.2.2. Constraints

Clearly these prevention measures represent additional constraints – and therefore extra costs – to designers and operators of launchers and satellites:

- Additional mass and increased complexity due to the extra equipment required for conducting passivation: valves, tubing, nozzles, pyrotechnics, etc.
- Inability to select the optimal injection orbit from a performance point of view in order to comply with the 25-year rule, need to re-ignite a launcher stage, etc.
- Shortening of the operational lifespan because of the mass of fuel needed to conduct end-of-life manoeuvres (and even more so because end-of-life uncertainties may require safety margins to be added to avoid running out of fuel).
- Cost of the operations themselves: teams of operators and specialists at the control centre, network of stations needed for the TM/TC link.
- Use of materials that do not generate debris when they age.
- Difficult decision-making with regard to terminating the mission in order to conduct end-of-life manoeuvres: the operator will wish to prolong the profitable use of its satellite as much as possible.

These prevention measures therefore require additional immediate expense on the part of the operators without any obvious benefit to them. Implementation of these measures has a long-term effect that concerns the entire community. Operators must therefore be encouraged (compelled) to apply these measures; however, because space activity is developed in a context of economic competition, each country, or each agency, cannot impose these (sometimes considerable) constraints if other competitors do not follow suit. For this reason, the issue needs to be debated on an international level and a consensus needs to be reached between all the stakeholders.

4.3.3.3. Regulatory provisions

For more than 15 years the issue of space debris has been raised by various authorities and numerous documents have been written. These include the following:

- The ITU (International Telecommunication Union): ITU-RS-1003 Recommendation on GEO disposal (re-orbiting geostationary satellites at the end of their mission).
- Space agencies: NASA standards first, CNES standards in 1999.
- Network of centres (ASI, BNSC, CNES, DLR and ESA): publication of the European Code of Conduct for Space Debris Mitigation in 2004.
- The IADC (Inter Agency Space Debris Coordination Committee): publication of IADC Space Debris Mitigation Guidelines in October 2002.
- The Scientific and Technical Sub-Committee (STSC) of COPUOS (Committee on the Peaceful Uses of Outer Space) published the UNCOPUOS Space Debris Mitigation Guidelines in 2007.
- Standards organisations such as ISO have developed standards on space debris.
- Countries have set up national regulatory provisions in the form of licence systems (USA, UK) or laws (France).

These documents have been developed to respond to immediate requirements and the situation may seem somewhat confusing to an outside observer bewildered by their number and respective roles.

Fortunately, despite having been written by different groups, these documents are technically consistent: the members of these groups almost all belong to the

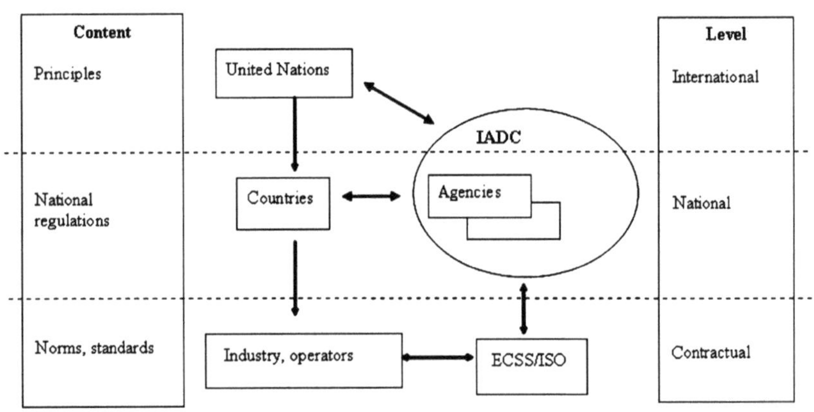

Fig. 11. *Summary of the situation: principles of the regulatory documentation (source: CNES).*

IADC and consistency has been maintained by the group of IADC experts who played a key role in the process.

Today the situation is clearer and can be summarised as follows (see Figure 11):

- The IADC is the common reference for all these documents. The IADC Space Debris Mitigation Guidelines represent the consensus of the 11 main space agencies: ASI (Italy), BNSC (United Kingdom), CNES (France), CNSA (China), DLR (Germany), ESA (Europe), ISRO (India), JAXA (Japan), NASA (USA), NSAU (Ukraine) and ROSCOSMOS (Russia). The IADC defines the technical bases for the measures (described in Section 3.2.1) and studies their influence on the environment.
- The United Nations, via COPUOS, defines the major principles for the measures to be applied in space: the seven principles are described in document A/AC.105/C.1/L.284, which was approved by the United Nations General Assembly in November 2007 (Resolution A/RES/62/217). The UN requests all States to implement these principles through national legislation to make them applicable. This adds to concerns about documents already drafted by the UN (Outer Space Treaty of 1967 and the Convention of 1972) that assign liability to the launch State in the event of damage caused in orbit or on the ground. This implies that the State must have the means to control the space activities of its own citizens.
- These States therefore need national regulations in order to impose measures on their operators and manufacturers to protect the orbital environment. Initially, the national space agencies, which conduct most of the space activities in their own countries, developed their own internal rules (NASA standard, CNES standard and European Code of Conduct). Nowadays, with space activities being increasingly carried out by the private sector, the agencies' self-imposed rules need to be replaced by laws applying to all a particular State's citizens: this is the role of the licence systems and national laws covering space activities.
- Lastly, operators and manufacturers need to translate these rules into implementation standards that can be directly applied to projects: these standards must be clearly measurable, quantifiable and verifiable. This corresponds to the work currently being undertaken within ISO.

4.3.4. Summary

The main space agencies, particularly NASA, were very quick to recognise the new problem posed by the proliferation of space debris. Because of their dual role as perpetrator and victim of this pollution, they reacted by developing internal rules applicable to their own projects and aimed at reducing the production of debris. It quickly became obvious that international consensus was also needed for these measures and this led to the creation of the IADC: this inter-agency committee is now the sole reference in the field and the space agencies all apply the same measures. Nowadays, however, the agencies are no longer the only stakeholders in space: an increasing share of activities is being conducted by private manufacturers and operators who can sidestep the agencies' rules. In accordance with the United Nations Treaties, liability in the event of any damage in space falls on States. The subject was therefore added to the agenda of the United Nations Committee responsible for space affairs (COPUOS), which has led to the drafting of a high-level document, still based on the IADC reference, that lays out the major principles to be applied in space with regard to prevention of space debris. These principles must then be passed on through each State's legislative system to ensure that the activities of all stakeholders in space can be controlled.[45]

[45] The following references were used for this article: Alby, Fernand. "Implementation of Space Debris Mitigation Guidelines." Proceedings of the Fifty-sixth International Astronautical Congress, 17–21 Oct. 2005, Fukuoka, Japan. IAC-05-B6.3.08; Alby, Fernand. "SPOT-1 end of life disposition manœuvres." Advances in Space Research 35.7 (2005): 1335–1342; Johnson, Nicholas L. "Current characteristics and trends of the tracked satellite population in the human spaceflight regime." Proceedings of the Fifty-seventh International Astronautical Congress, 2–6 Oct. 2006, Valencia, Spain. IAC-06-B6.1.03.

4.4 Space situational awareness: an overview

Giovanni Gasparini and Valérie Miranda

4.4.1. Introduction

In the past 20 years we have experienced rapid growth in the launch of space assets by dozens of nations and operators, mostly for communications and entertainment (TV) purposes but also to provide increased services for the public in general (navigation systems for example) and for the military in particular. Hundreds of satellites, some of which are no longer operating, are crowding the most popular orbits (GEO, geostationary orbit, and some specific LEOs – low-Earth orbits), accompanied by thousands of debris travelling at speeds of kilometers per second, enough to transform even a very small (1 cm) piece of debris into a killer bullet for thin protective skins. It was only a matter of time before this unregulated system, recently put under further strain by the Chinese anti-satellite weapons (ASAT) test of 11th January 2007 that generated thousands of debris still remaining in orbit, was shaken by a potentially catastrophic event.[46] It eventually happened on 10 February 2009 when two satellites collided creating a "cloud" of thousands of debris potentially producing a chain effect along one of the most crowded LEO orbits. The impact was unforeseen despite both satellites being present in the U.S. space catalogue. If the commercial operator of one of the satellites (Iridium, which provides communications for the wider public and the Pentagon) had received timely information, it could have manoeuvred in order to avoid the dead Russian military satellite into which it crashed. This accident shows how little timely information we have as to the traffic in orbit and how important it is in economic, operational and strategic terms to improve the current situation. This article offers a general overview of the current situation regarding the availability of systems providing this precious information, focusing on the strategic value of space awareness systems, and discusses the implications of different approaches towards the establishment of a wider global SSA system.

4.4.2. Present SSA situation

The term "Space Situational Awareness" was coined by the U.S. Air Force. During the Second World War, the German and Allied Air Forces observed that most of the fighter pilots hit did not realise they were under enemy fire until their plane was destroyed. Similarly, later studies conducted by the U.S. Air Force (USAF) concluded that during the war in Korea and Vietnam, 80% of planes that were shot were completely unaware of what was happening. According to USAF experts, the main cause was a lack of "Situation Awareness" (SA), which thus proved essential to the fighter pilots' survival.[47] The concept of Situational Awareness was then used to refer to the space environment and appeared as such for the first time in the 2001 Report on Space by Donald Rumsfeld.[48]

As Laurence Nardon clearly explains in her "Space Situational Awareness and International Policy", SSA – in plain words, the ability to "see" what is going on in space – "can have different applications and can therefore serve different policy goals". In particular, "in the United States, SSA programmes monitor the threat from human-made objects: other satellites and space vessels, anti-satellite weapons (ASATs) as well as space debris".[49] Thus, as well stated by Theresa Hitchens, "SSA is the foundation stone underpinning all operations in space, required for ensuring that working satellites do not interfere with each other, debris tracking and collision avoidance, diagnosing an ailing satellite, and satellite protection and defence,[50] as well as for the more controversial Air Force mission of "offensive space control".[51] In the European context, however, SSA is defined in much wider terms, including 'the awareness of threats from asteroids, solar flares and other "astronomical threats"'.[52]

4.4.2.1. SSA in the United States

The United States is so far the only country that has developed a global SSA system,[53] the "Space Surveillance Network" (SSN). This is composed of ground-based radars and optical sensors located in 25 sites in the Northern hemisphere.[54] These sensors are grouped into three main categories: dedicated sensors, whose primary mission is space surveillance and which are owned by the Air Force Space Command; collateral sensors, initially conceived for missile warning and now used for space surveillance missions; and contributing sensors which provide data as part of the SSN but are owned by private contractors or by other branches of the U.S. government.

The SSN tracks space objects which are 10 cm in diameter or larger in both LEO and the higher geostationary orbit at an altitude of 36,000 km where telecommunications satellites operate. The space objects tracked consist of active and inactive satellites, spent rocket bodies and fragmentation.[55]

The enormous amount of data collected by SSN sensors[56] is then transmitted via satellite, ground wire, microwave and telephone to the Joint Space Operations Centre (JSpOC), which is part of the United States Strategic Command (USSTRATCOM). The JSpOC then fuses the SSN data with other sources to provide SSA for the U.S. military and other customers.[57]

In addition, data are regularly published and used free of charge worldwide by different users interested in tracking satellites and space debris.[58] In this regard it is important to underline that while published U.S. information includes data on the orbits of other nations' military hardware, it excludes data on sensitive U.S. defence satellites.[59]

Although it is a reference point for cataloguing satellites and space debris, the U.S. SSN has some limitations that are acknowledged by field experts and Department of Defence (DoD) officers. The former complain of low accuracy, incomplete information and a proprietary data format (the so-called two lines element);[60] the latter recognise that "U.S. SSA capabilities are less than adequate today [. . .] the sensors cannot consistently find small debris and have limited capability to find, track and characterise objects in high-altitude orbits".[61] It is therefore generally admitted that the U.S. should devote more resources to the enhancement of its SSA capabilities. This is particularly true if we consider the growing importance of space assets and the increasing U.S. reliance and dependency on them. Indeed, from Desert Storm to recent operations in Afghanistan and in Iraq, military interventions have increasingly depended on space capabilities as force multipliers. Furthermore, space assets have become essential to worldwide commerce and everyday life. To date, American supremacy in the space environment is still unchallenged but this is likely to change in the future and the proliferation of space capabilities as well as the emergence of new competitors (such as China, or Europe in the satellite navigation field) has to be taken into account.[62]

So far, U.S. space policy has focused on Space Force Enhancement (SFE), which provides combat support operations to improve the effectiveness military forces, and on Space Force Application (SFA)[63] that is mostly devoted to nuclear deterrence. However, considering the emerging threats to its space assets and the consequent need to ensure its control of space as well as space superiority, U.S. attention is now shifting towards the enhancement of Counterspace (CS) activities. These latter are defined as the "capabilities needed to attain and maintain a desired degree of space superiority by allowing friendly forces to exploit space capabilities while negating an adversary to do the same".[64] They are made up of three "pillars": SSA, Defensive Counterspace (DCS) and Offensive Counterspace (OCS).

Within the general framework of U.S. National Space Policy that was adopted by the Bush Administration in 2006, the Strategic Master Plan (SMP) of the Air Force Space Command well illustrates the three-phase process to enhance U.S. Counterspace capabilities over the next 15 years.

First, with respect to SSA which is considered to be the "permanent crucial enabler" for DCS and OCS, the SMP recognises the need to improve U.S. ability to "find, fix, track and provide characterisation data on near earth and deep space objects and events, as well as improving the ability to adequately process and analyse data from all space regimes and from all SSA sources".[65] Additionally, the U.S. should enhance its ability to distinguish man-made attack and other sources of anomalies from natural environmental effects. Finally, the SMP provides for the current SSN system to be modernised and complemented with various in-orbit telescopes as well as innovative detection systems on board future satellites.[66]

Second, as concerns Defensive Counterspace, in addition to current DCS techniques that now focus on hardening satellites against electronic jamming, further means (in-orbit manoeuvring capabilities, launch-on-demand capacities, satellites redundancy and smallsats constellations) should be available after 2018.[67]

Third, with regard to OCS capabilities, the aim is to negate adversary space services by creating reversible (deceive, deny, disrupt) or irreversible (degrade, destroy) effects. According to the Strategic Master Plan, these are the least urgent capability, to be dealt with only in the long term. Thus the Plan provides that by 2025 a range of means, including lasers and in-orbit ASATs, will be added to the electromagnetic pulse (EMP) jamming systems currently available, targeting all existing satellite systems.[68]

Overall, it seems that Space Situational Awareness in the U.S. serves mainly military purposes. However, there is no general consensus as to its specific use or final scope. According to some experts, SSA could be one of the measures at U.S. disposal for deterring future attacks against its satellites. As John Sheldon puts it, "effective deterrence is strengthened by the fact that Space Situational Awareness could potentially indicate the nature and origins of any attempted attack on a satellite".[69] In contrast, others such as Robert Butterworth observe that mere deterrent policies are not effective enough to reduce the vulnerability of U.S. space assets and recommend relying on defence measures: "defence can deter, but deterrence policies cannot defend. Defence can be tested and exercised; deterrence threats cannot: their efficacy depends on the perceptions and actions of a foreign government."[70] Butterworth further states that: "none of these (defence) measures requires anything in the way of space-based weaponry".[71]

The debate over the best way to protect American space capabilities and ensure U.S. control of space is thus part of a wider and heated discussion over space weaponisation in which SSA initiatives play a key role. Indeed, some argue that these latter are the first step towards the acquisition of space-based weaponry. Laurence Nardon, for instance, claims that while the Eisenhower Administration formally excluded the weaponisation of space in 1958, deeming it to be too destabilising, the 2001 Rumsfeld report and the three-phase-USAF plan represent a change in attitude. The 2006 National Space Policy seems to go in the same direction. Despite the denials of the Bush Administration, the principle that "the United States will oppose the development of new legal regimes or other restrictions that seek to prohibit or limit U.S. access to or use of space"[72] has been interpreted by many experts as a thinly veiled authorisation of space weaponisation. For instance, Michael Krepon, co-founder of the Henry L. Stimson Center, said this new policy would reinforce international suspicions that the United States may seek to develop, test and deploy space weapons.[73]

However, it seems that this might change under the new U.S. President, Barack Obama. In fact, after Obama's inauguration, the official White House web site was updated with new policy guidelines including one on restoring U.S. leadership in space which confirmed the goals already outlined in Obama's 2008 campaign.[74] Under the heading "Ensure Freedom of Space", US official policy is to "seek a worldwide ban on weapons that interfere with military and commercial satellites; assess possible threats to U.S. space assets and the best military and diplomatic means for countering them; seek to assure U.S. access to space-based capabilities, in part by accelerating programmes to harden U.S. satellites against attack".[75] While experts generally agree that Obama's intentions could lead to a new direction in space diplomacy, most of them are waiting for further specific details such as the definition of the controversial term "space weapon".[76]

Even if tough SSA military implications are prevalent, supporters of the non-weaponisation of space look at the other side of the coin. Indeed, agreeing on the need for better awareness of what happens in space, they adopt a different perspective and consider SSA "as a major tool to enable a continuing peaceful use of space".[77] To this end, a specific proposal is that of Brian Weeden, Secure World Foundation, who recommends the creation of an international civil Space Situational Awareness system whose goal would be to "provide all space actors access to the tools needed for safe and sustainable activity in Earth orbit".[78] The fundamental difference between this kind of system and military SSA is "in the information it provides, focusing only on the locating of an object in Earth orbit and a point of contact for that object, along with information about space weather".[79] Moreover, such a civil system could provide several benefits to the international community. In addition to the traditional information generally provided by SSA systems, Weeden says it could increase international cooperation and transparency (and therefore mutual trust) in space activities and also be a potential verification mechanism for a code of conduct or a space traffic management system that might be created in the future.

4.4.2.2. SSA in Europe

As mentioned above, the European definition of Space Situational Awareness is wider than that used in the U.S. According to the "User Expert Group of ESA SSA Requirement Study", Space Situational Awareness can be defined as "the comprehensive knowledge of the population of space objects, of existing threats and risks, and of the space environment".[80] The term "Space Surveillance" refers, "instead, to the routine operational service of detection, correlation, characterisation and orbit determination of space objects".[81] A certain overlap between the two concepts therefore exists, but "SSA implies more in terms of data processing and use".[82]

The importance that space assets have for U.S. policy is also confirmed in the European context. Indeed, according to the 2007 European Space Policy, space is a strategic tool for independence, prosperity, development and progress from an economic, technological, scientific and societal point of view. Moreover, the use of space assets has become essential, especially for security and defence purposes.[83] To this end, European countries and the EU have developed important space infrastructure (for communications, positioning, monitoring and intelligence applications) that needs to be operated safely to prevent possible collision, disruption or malfunction.[84] SSA activities are crucial in this regard. However, as pointed out in the "Conclusions of the Workshop on Space Security and the

Fig. 12. *GRAVES Space Surveillance System: receiving antennas (source: ONERA).*

Role of the EU" held in Berlin in June 2007[85], Europe does not have an autonomous capability for SSA: some sensors exist at the national level (Figure 12) but they can provide only part of the necessary information. Thus, Europe depends heavily on data provided abroad, namely by the U.S. SSN.

In order to fill this gap and in accordance with the aforementioned Conclusions, Europe has been pursuing plans to create its own SSA assets focusing in particular on the monitoring of space and on the identification of potential natural and man-made threats to its security. In 2006, ESA created a task force on space surveillance and began a series of parallel studies aimed at identifying SSA end-users needs to translate them into technical requirements. Furthermore, in the framework of preparatory action, ESA commissioned a number of studies on possible SSA architecture, hardware, governance and data policy.[86] These initiatives led to a recent decision of the ESA Council at Ministerial level that in November 2008 finally endorsed the SSA preparatory programme.[87] The original idea was to launch a 5-year full programme with a budget of 100 million euros. However, due to the political concerns of Member States, together with changing perspectives related to space weather or near-Earth objects and the general financial and economic crisis, the proposal was then changed to a preparatory and optional programme with a budget of 50 million euros[88] and a time frame of 3 years (2009–2011).[89] As reported by Wolfgang Rathgeber, according to the proposal the programme is composed of a core element and three additional optional elements. The first includes issues that are of concern to most ESA Member States such as governance, data policy and data security as well as architecture and space surveillance. The three optional elements concern space weather and near-Earth objects, breadboarding of radar components and pilot data centres. Practically, the programme largely consists of studies and workshops related particularly to governance and data policy issues, whereas questions referring to space surveillance, space weather and near-Earth objects will be addressed separately. Furthermore, as soon as the data policy section is completed, some pre-cursor SSA services will be established relying on pre-existing national facilities. Hardware activities, such as the above-mentioned breadboarding of radar components together with an initial procurement of assets and components, have been planned.[90]

The ESA preparatory programme is certainly an important step towards the creation of future European autonomous SSA capabilities. Nevertheless, other concrete improvements are needed. In view of its preparatory nature, the ESA programme does not contain suggestions how to concretely realise the future European SSA system. In this respect, some experts suggest that a European SSA system would initially collect data from existing national level assets, mostly ground-based national sensors.[91] But it is likely that this system would gradually change over time, relying more and more on European rather than national assets.

For this reason, its architecture should be flexible enough to adapt to progressive integration at the European level.[92]

A good starting point for a future SSA system could be along the lines of GRAVES, the French national radar system owned and operated by the French Air Force to keep under surveillance and track space objects in LEO. However, considering its predominantly military use together with its national and geographic limits, its possible influence on a European SSA system should be carefully assessed.[93]

The European SSA system's possible future architecture is not the only aspect to be dealt with in future discussions. These should focus on (and find convincing solutions for) two main critical issues: the need to reconcile different end-users' needs and the need to elaborate a mechanism to manage the amount of data coming from national assets in an effective and coordinated way. As regards the first aspect, the European SSA system is generally conceived to be a dual-use system and to provide services for four categories of end-users (institutional, military, commercial and scientific ones) with different needs in terms of security requirements, governance and data policy. The most demanding end-users are military users who require relevant, reactive and precise information together with protection of the confidentiality of security and defence-related data. In this regard, some wonder whether ESA is the appropriate body to handle this kind of data. In fact it seems that military entities are not inclined to deal with security issues in a transparent body such as ESA, deciding instead to mandate the European Defence Agency (EDA) to consolidate the military requirements for an SSA system. Thus the question is how to coordinate civilian and military aspects of security as well as the relationship between ESA and EDA. Recommendations in this regard will be presented at the next ESA Ministerial Council in 2011.[94]

With reference to the second aspect, governance and management of data, a generally acknowledged need is to reconcile the possible coexistence of different data policies in a coordinated and flexible way. Concrete proposals in this regard concern the creation of operational schemes and data exchange formats able to: ensure the optimum availability of information (including necessary redundancy and the avoidance of unnecessary duplication), provide adequate incentives for all potential contributors and, last but not least, take account of military bodies' security concerns and the economic interests of commercial participants.[95]

Europe is an important space actor but, as Nicolas Bobrinsky (ESA) has insightfully said, Europe "has, right now, little to offer".[96] It is generally acknowledged that the development of autonomous SSA capabilities will have important positive effects. SSA would increase European knowledge and understanding of the global situation in space and better prepare Europe to react to any risk posed by

Fig. 13. *Tira System (source: FGAN).*

the loss of satellite or related services or by a collision with an asteroid or a comet.[97] This would also lead to benefits from the strategic point of view since it is likely that the European position towards major space partners, such as the U.S., would be strengthened thus most likely filling in some of the current deep gaps between the two.[98]

4.4.3. Defence, deterrence and SSA

As the above discussion of the complex debate surrounding SSA shows, the U.S. model, based on the primacy of the defence dimension, is predominant. In view of the heavy dependence of the U.S. military on space force multipliers for both conventional military operations and strategic nuclear policy, concerns in this area are well-grounded.

The relative weakness of space assets (as was further demonstrated by the American use of a modified Missile Defence interceptor and related assets in order to destroy a rogue U.S. intelligence satellite in 2008), makes the U.S. an attractive target for an asymmetrical attack (weak to strong). This is particularly true if we bear in mind that in the future an increasing number of countries will have access to space (while at the same time being less reliant on space than the U.S. military) and, due to the absence of a clear identification system, could launch an attack without being identified or held accountable for it.

In this respect, the space environment could look similar to the cybersphere where the current difficulty of tracking the origin of an attack nullifies potential law enforcement or dissuasion responses. In order to re-establish deterrence, it is necessary to field a reliable system that determines the origin of a potential attack against a satellite and makes it possible to manoeuvre to counter the attack and retaliate against it. The system should also be able to avoid false alarms and to distinguish between deliberate attacks and accidental interference.

Attacking a military asset in space is an act of war that carries all the political, legal and operational consequences that apply to ground attacks. Adopting a deterrence policy that clearly states the will and intent to react in a tit-for-tat fashion, not only against another space asset but for example against land-based space facilities that give access to space, would be proportional and stabilising. The availability of a reliable SSA system is essential to establish the credibility of such a deterrent.

A second element of the defence-based model concerns secrecy of information. In the defence realm, where a zero-sum game often prevails, asymmetry is a positive result; therefore, situational knowledge of space gives additional power to the owner of the information. Moreover, in order to be effective, data-collectors, classified intelligence and observation satellites need the highest possible level of secrecy concerning their orbits, characteristics and capabilities.

The situation described above would naturally lead towards the fielding of a number of separate independent national space situational awareness systems. However, such a solution has important shortfalls. The development of a reliable and accurate system is feasible for possibly only one player – the U.S. If they managed to combine their efforts, European countries would also have this potential but to a lower extent. This, however, would have the effect of reducing the confidence of other space powers in national systems (Europeans already do not consider the present U.S. system reliable for both technical and political/access reasons), thus reducing to zero the confidence-building effect of fielding an SSA.

This over-militarised vision of an SSA is not having a positive impact on commercial operators and other users, thus potentially leaving out of the equation a number of stakeholders whose cooperation would be very important for the actual success of a system that aims at avoiding collisions.

4.4.4. SSA as a dual strategic asset

The European approach seems to be taking a different direction. Commercial and institutional non-military assets outnumber dedicated military satellites. Compared to the U.S., European military forces rely less on space assets, with some exceptions regarding France where some space assets are performing conventional military, intelligence and nuclear-related missions. The United Kingdom is noticeably silent in this discussion as its intelligence and command structure relies heavily on the tacit availability of U.S. assets.

The SSA mandate given to ESA by member countries is consistent with the "peaceful purposes" of the organisation. A superficial analysis would suggest that ESA is not sufficiently taking into account the requirements of military operators. However, particularly at the beginning, the European SSA will be the result of the merging of national assets and data in which national governments will exercise strong indirect control, making sure that "sensitive" information will not be shared widely.

Transparency is a paramount principle guiding European efforts, as well as openness to commercial and scientific operators. While recognising the intrinsic dual (civil/military) role of any SSA system, the ESA is reluctantly discussing the security implications of wider availability of data that potentially also applies to classified military observation satellites.

However, national authorities operating in space and security, particularly France, are well aware of the strategic value of controlling space assets. It is no secret that the national French GRAVES system aims to control which foreign (in particular American) intelligence-gathering satellites are actually over-flying France, thus obtaining a bargaining chip in order to convince U.S. authorities to stop publishing sensitive data concerning similar French satellites.[99]

The European Space Agency is rightly seeking to convince its Member States that the few national efforts in this field are providing limited results and offering to move a step forward towards a federated system. Such a system would then constitute the European contribution to a global structure involving information from all willing space nations and commercial operators.

This approach is less concerned with voluntary actions that purposefully interfere with satellites. Instead, it encompasses a series of potential applications regarding the space environment and space traffic where the characterisation of the space object is less important than in the case of military applications while remaining still relevant for determining legal liabilities.

However, as the number of nationally owned and common European space assets is bound to increase in the next few years pursuant to programmes having strong security implications such as Galileo, GMES and Musis, the Europeans will increasingly be forced to think strategically and the dual character of the SSA system will emerge strongly.

4.4.5. Reconciling different approaches in the international arena

A Space Awareness system that does not work as a confidence-building measure between potentially competing space actors will inevitably increase the likelihood of a conflictual posture in space that exploits the asymmetrical vulnerability of U.S. military space assets. This could also create an environment in which non-military security and commercial satellites would not be adequately protected.

A commonly agreed governance and data policy system that resolves the trade-off between the effectiveness of the transparency approach and the secrecy requirements of the military and intelligence community could bridge the current gap between the European and U.S. positions. The key to this approach is to allow differentiated access to data according to the real "need to know" of the potential users. In the case of commercial operators and the wider public, this would exclude knowledge of the characterisation of satellites unless specifically requested when an event requiring the assessment of legal liability occurs. U.S. authorities need to take more account of the dual character of space. At the same time, European institutions need to think more strategically. This discussion should take place between all U.S. Space Agencies on the one side and the European Council, the European Commission and ESA on the other.

The problem with other space nations that are not bound by the Transatlantic Alliance is however much more complicated. China and to a lesser extent Russia, as well as other minor space-capable countries such as Iran, would feel potentially threatened by a non-inclusive American or even transatlantic approach to space awareness. As it is unlikely that they will field a national SSA system, the incentive for them to develop ASAT capabilities would be high. This is particularly true due to the complexity and high cost of defending a space asset compared to the relative small cost of attacking it.

However, it has to be made clear that access to global SSA information will imply the acceptance of rules concerning contributions to a common database, the use of data and general behaviour in space. Cooperation cannot be seen as a way of free-riding or, worse, exploiting common knowledge for illicit purposes such as targeting of space assets. Ideally, the effort would include a common set of rules and possibly a treaty re-establishing an ASAT ban. Unfortunately that would be difficult due to the potential use of missile defence systems for that purpose. It will not be easy to strike the right balance between the different needs of nations and users, but it is necessary to reach it soon as the space community cannot afford that further casual or deliberate clashes occur.

[46] For further reference, see Secure World Foundation. "Chinese Anti-Satellite (ASAT) Test." 6 June 2008. SWF Factsheet. 9 Mar. 2009. http://www.secureworldfoundation.com.
[47] Nardon, Laurence. "Space Situational Awareness and International Policy." Oct. 2007. Document de travail 14. Institut Français des Relations Internationales. 15 June 2009. http://www.ifri.org/files/Espace/Docu_14_SSA_Nardon.pdf.
[48] "Report of the Commission to Assess U.S. National Security Space Management and Organization." 11 Jan. 2001. Executive Summary. Pursuant to public Law 106–65. 15 June 2009. http://www.fas.org/spp/military/commission/executive_summary.pdf.
[49] Nardon, Laurence. "Space Situational Awareness and International Policy." Oct. 2007. Document de travail 14. Institut Français des Relations Internationales. 15 June 2009. http://www.ifri.org/files/Espace/Docu_14_SSA_Nardon.pdf: 2.
[50] For a more specific list of space surveillance purposes, see Air University Website. http://www.au.af.mil/au/awc/awcgate/usspc-fs/space.htm.Last accessed 9 Mar. 2009.
[51] Hitchens, Theresa. "Ante up on Space Situational Awareness." Space News 12 Mar. 2007: 19.
[52] Nardon, Laurence. "Space Situational Awareness and International Policy." Oct. 2007. Document de travail 14. Institut Français des Relations Internationales. 15 June 2009. http://www.ifri.org/files/Espace/Docu_14_SSA_Nardon.pdf: 2.
[53] Marta, Lucia and Giovanni Gasparini. "Europe's Approach to Space Situational Awareness: a Proposal." Yearbook on Space Policy 2007/2008: From Policies to Programmes. Eds. Kai-Uwe Schrogl, Charlotte Mathieu and Nicolas Peter. Vienna: SpringerWienNewYork, 2009: 1.
[54] Such as Maui, Hawaii and Eglin, Florida; Thule, Greenland and Diego Garcia, Indian Ocean.
[55] Of the space objects tracked, 7% are operational satellites (in which the US is most interested), 15% are rocket bodies and 78% are fragmentation and inactive satellites. http://www.au.af.mil/au/awc/awcgate/usspc-fs/space.htm. Last accessed 9 Mar. 2009.
[56] The SSN has been tracking space objects since 1957 when the Soviets opened the space age with the launch of Sputnik I. Since then, 24,500 space objects orbiting Earth have been tracked. http://www.au.af.mil/au/awc/awcgate/usspc-fs/space.htm. Last accessed 15 June 2009.
[57] For more details, see Secure World Foundation. "Space Situational Awareness." 10 June 2008. SWF Factsheet. 9 Mar. 2009. http://www.secureworldfoundation.com.
[58] The catalogue is available publicly at http://space-track.org. Last accessed 16 June 2009.
[59] de Selding, Peter B. "French say "Non" to US Disclosure of Secret Satellites." 8 June 2007. Space.com. 9 Mar. 2009. http://www.space.com.
[60] Marta, Lucia and Giovanni Gasparini. "Europe's Approach to Space Situational Awareness: a Proposal." Yearbook on Space Policy 2007/2008: From Policies to Programmes. Eds. Kai-Uwe Schrogl, Charlotte Mathieu and Nicolas Peter. Vienna: SpringerWienNewYork, 2009: 11.
[61] U.S. Air Force Space Command. Strategic Master Plan FY06 and Beyond. 1 Oct. 2003. 15 June 2009. http://www.wslfweb.org/docs/Final%2006%20SMP–Signed!v1.pdf: 22.
[62] Ibid.: 4.
[63] SFA is defined as the "capabilities to execute missions with weapons systems operating from or through space which hold terrestrial target at risk", U.S. Air Force Space Command. Strategic Master Plan FY06 and Beyond. 1 Oct. 2003. 15 June 2009. http://www.wslfweb.org/docs/Final%2006%20SMP–Signed!v1.pdf: 2.
[64] U.S. Air Force Space Command. Strategic Master Plan FY06 and Beyond. 1 Oct. 2003. 15 June 2009. http://www.wslfweb.org/docs/Final%2006%20SMP–Signed!v1.pdf: 2.
[65] Ibid.: 22.
[66] Nardon, Laurence. "Space Situational Awareness and International Policy. " Oct. 2007. Document de travail 14. Institut Français des Relations Internationales. 15 June 2009. http://www.ifri.org/files/Espace/Docu_14_SSA_Nardon.pdf: 2.
[67] Ibid.: 2.
[68] Ibid.: 3.

[69] Sheldon, John B. "Space Power and Deterrence: Are We Serious?" Nov. 2008. Policy Outlook. George Marshall Institute, Washington. 15 June 2009. http://www.marshall.org/pdf/materials/616.pdf.

[70] Butterworth, Robert. "Fight for Space Assets, Don't Just Deter." Nov. 2008. Policy Outlook. George Marshall Institute, Washington. 15 June 2008. http://www.marshall.org/pdf/materials/614.pdf.

[71] Ibid.

[72] U.S. National Space Policy. 31. Aug. 2006.

[73] Kaufman, Marc. "Bush Sets Defense as Space Priority." 18 Oct. 2006. The Washington Post 15 June 2009. http://www.washingtonpost.com/wp-dyn/content/article/2006/10/17/AR2006101701484.html.

[74] See http://www.whitehouse.gov/agenda/defense Last accessed 10 Mar. 2009; "Advancing the Frontiers of Space Exploration" http://www.barackobama.com/pdf/policy/Space_Fact_Sheet_FINAL.pdf. Last accessed 16 June 2009.

[75] Brinton, Turner. "Obama Space-Weapon Ban Draws Mixed Response." FoxNews.com. 16 June 2009. http://www.foxnews.com/story.

[76] Ibid.

[77] Nardon, Laurence. "Space Situational Awareness and International Policy." Oct. 2007. Document de travail 14. Institut Français des Relations Internationales. 15 June 2009. http://www.ifri.org/files/Espace/Docu_14_SSA_Nardon.pdf: 3.

[78] Weeden, Brian. "Notes on Civil SSA." Presentation. 46th Session of Scientific and Technical Subcommittee of the United Nations Committee on the Peaceful Uses of Outer Space (COPUOS). Vienna, Austria. 17 Feb. 2009. 16 June 2009. http://www.secureworldfoundation.org/siteadmin/images/files/file_277.pdf.

[79] Ibid.

[80] User Expert Group of ESA SSA requirement study, available at:www.sidc.be/esww4/presentations/SWWT/SSA%20-%20Space%20Weather%20Week.ppt. Last accessed 16 June 2009.

[81] Task force on Space Surveillance BNSC, CNES, DLR, ESA, available at: www.sidc.be/esww4/presentations/SWWT/SSA%20-%20Space%20Weather%20Week.ppt. Last accessed 16 June 2009.

[82] User Expert Group of ESA SSA requirement study, available at: www.sidc.be/esww4/presentations/SWWT/SSA%20-%20Space%20Weather%20Week.ppt. Last accessed 16 June 2009.

[83] Marta, Lucia and Giovanni Gasparini. "Europe's Approach to Space Situational Awareness: a Proposal." Yearbook on Space Policy 2007/2008: From Policies to Programmes. Eds.Kai-Uwe Schrogl, Charlotte Mathieu and Nicolas Peter. Vienna: SpringerWienNewYork, 2009: 14; Council of the European Union. Resolution on the European Space Policy. Doc. 10037/07 of 22 May 2007. Brussels: European Union.

[84] Marta, Lucia and Giovanni Gasparini. "Europe's Approach to Space Situational Awareness: a Proposal." Yearbook on Space Policy 2007/2008: From Policies to Programmes. Eds.Kai-Uwe Schrogl, Charlotte Mathieu and Nicolas Peter. Vienna: SpringerWienNewYork, 2009: 1.

[85] EU-Conference on "Security in Space, the Contribution of Arms Control and the Role of the EU", Berlin, 21st–22nd June 2007; Conclusions of the Workshop on space security and the role of the EU, available at: www.sidc.be/esww4/presentations/SWWT/SSA%20-%20Space%20Weather%20Week.ppt. Last accessed 16 June 2009.

[86] Rathgeber, Wolfgang. "Space Situational Awareness (SSA) for Europe, a First Important Step." ESPI Perspective 16, December 2008. 16 June 2009. http://www.espi.or.at/images/stories/dokumente/Perspectives/espi%20perspective%2016.pdf: 1.

[87] ESA. European Space Agency Ministerial Council. The Hague, 25–26 Nov. 2008. Final Conclusions. Doc. ESA PR 44-2008. Paris: ESA.

[88] Eleven Member States subscribed the programme and Spain is going to be the main contributor.

[89] ESA. European Space Agency Ministerial Council. The Hague, 25–26 Nov. 2008. Final Conclusions. Doc. ESA PR 44-2008. Paris: ESA.

[90] Rathgeber, Wolfgang. "Space Situational Awareness (SSA) for Europe, a First Important Step." ESPI Perspective 16, December 2008. 16 June 2009. http://www.espi.or.at/images/stories/dokumente/Perspectives/espi%20perspective%2016.pdf: 2.
[91] Marta, Lucia and Giovanni Gasparini. "Europe's Approach to Space Situational Awareness: a Proposal." Yearbook on Space Policy 2007/2008: From Policies to Programmes. Kai-Uwe Schrogl, Charlotte Mathieu and Nicolas Peter (Eds). Vienna: SpringerWienNewYork, 2009: 7–8.
[92] Ibid.:7–8.
[93] Ibid.: 5.
[94] Rathgeber, Wolfgang. "Space Situational Awareness (SSA) for Europe, a first important step." ESPI Perspective 16, December 2008. 16 June 2009. http://www.espi.or.at/images/stories/dokumente/Perspectives/espi%20perspective%2016.pdf: 2–3.
[95] Ibid.: 2–3.
[96] Interview with Nicolas Bobrinski, Head of ESA Ground Station System Division. 13 Nov. 2008. available at: http://www.esa.int/SPECIALS/Space_Debris/SEMFSG6EJLF_0_iv.html, last accessed 9 Mar. 2009.
[97] Ibid.
[98] Nardon, Laurence. "Space Situational Awareness and International Policy." Oct. 2007. Document de travail 14. Institut Français des Relations Internationales. 15 June 2009. http://www.ifri.org/files/Espace/Docu_14_SSA_Nardon.pdf: 5.
[99] de Selding, Peter. "France Pressures U.S. to Stop Publishing Orbits of French Milsats." 13 June 2007. Space.com. 9 Mar. 2009. http://www.space.com.

5 Fair rules on orbit – summarising introduction

Richard J. Tremayne-Smith

There were four speakers in this session covering a selection of issues related to what can be considered fair rules for use on orbit as well as proposals on extending and enhancing the status quo. The lively discussion following the presentations extended the scope of consideration with regard to fair rules that could or should apply both to activities on orbit and to those necessary to reach or transition through those orbits.

The session started with an interesting presentation by Wolfgang Rathgeber of ESPI on "Data Sharing Issues". The limitations of current systems and examples of different architectures that are or have been in use for space were considered. Then some data transfer examples were presented and finally an assessment of some current concerns that lead to proposals on the way forward. It was considered that collective data acquisition using new purpose built systems could lead to improved data sharing. The development of international collaborative activities in space weather could be used as a test case for improved data sharing. A complete data policy for sharing important data would need to consider not just the technical and system architecture issues but the legal aspects, such as those arising from the Remote Sensing Principles (applicable to Earth observation systems), as well as policy and political issues. More generally it was considered that joint missions and co-funding lead to a natural sharing of data but commercial missions can naturally lead to open data policies in order to achieve their marketing goals.

The second presentation in the session was given by Lubos Perek of the Czech Republic on the matter of "Cooperation within the UN System". The various UN bodies started collaborating on space issues in the 1970s. The main focus has been on practical applications of space, covered by the UN Office for Outer Space Affairs space applications programme, and work on mitigating natural disasters together with managing the relief effort. The presentation went on to focus on some specific issues related to the fair and responsible use of space including managing space traffic, and the transition between airspace and outer space, as well as looking at registration practices and the problems related to different organisations (ITU[100] and UN OOSA) registering respectively "networks" (or broadcasting positions and frequencies in space) and satellites. Problems were also seen to arise where some countries/users do not require prior ITU approval before allowing a new satellite to be launched or re-positioned. It was proposed that the solution could lie with a space equivalent to the civil aircraft control regime (ICAO[101]) that would provide more coordinated regulation and extend registration through rules of the road to space traffic management. It was proposed during discussion that, for now, States licensing the use of space should ensure satellite registration and general compliance with the Outer Space Treaties and should also include formal checks with the ITU both at initial license screening and before approval for launch.

The third presentation was by Ben Baseley-Walker of the Secure World Foundation on the status of work on current international space security issues, e.g., UN COPUOS[102] and the PAROS group of the Committee on Disarmament as well as reference to the work of the EU "Paris Group". The top-down issues were considered to arise from the Russia/China proposals (to PAROS) for no weapons in space and those focussed on an ICAO equivalent for space. The bottom-up issues were considered as the EU Code of Conduct from the "Paris Group", the informal working group within COPUOS and the current coordination amongst major satellite operators on situational awareness and space asset management. Overall, while the need is for full multilateral implementation of all possible confidence-building measures covering space debris, traffic management, etc., the practical route to implementation will need national, regional, civil and commercial activity to do whatever can be done as soon as reasonably possible. The key to success will be to show the various initiatives as part of a concerted effort to achieve fair and equitable use of space for all potential users of space. This will require caring for the space environment and ensuring it remains open for all humankind to use in the future without hindrance. While details of the EU initiative are not yet public it is expected that the proposals will include a wide range of confidence-building measures.

The final presentation in the fair rules on orbit session was by William Ailor of the Aerospace Corporation on "Space Traffic Control – issues and options". The orbital debris environment was described in some detail in order to explain the projected growth in the Earth orbiting debris and thus the ever-increasing risk of collisions with active spacecraft. The need for further data to help understand and manage the near-Earth environment was explained and the various players introduced. While tools have been developed to help manage space traffic the availability of good data is still an issue. The commercial players have data for their own satellites that can be better than the general-purpose government surveillance data. While there are clearly issues of trust, liability and availability due to security and commercial confidentiality issues, greater sharing of data and the results of the many supporting tools, such as those for collision avoidance, is essential. The presentation put "fair and responsible" in a practical context and was a good conclusion to the session on fair rules in orbit.

General conclusions from the discussion following the presentations were that fair rules on orbit should already be interpreted more generally (where complexity allows) to include not just Earth orbit but outer space in general and other special regions of outer space such as the Moon, Mars and various Lagrangian points with significant scientific and potentially commercial interest. Some delegates considered that many people do not understand some of the basics of orbital mechanics and where they apply, e.g., when ballistic trajectories in the atmosphere change so that more complex interactions need to be considered. Education and training will be an important part of the ongoing work to improve the understanding and facilitate the implementation of the growing number of supportive measures.

A simple and general plan for looking after space should be developed so that continued access, which is both fair and responsible, can be maintained. We have the Space Treaties and Principles, Guidelines, Standards and Codes of Conduct but what we also need is harmonisation and common implementation if the responsibilities embodied in these texts are to be truly fair. The results of the many proposed activities cannot be fully integrated today but it is essential we keep the full range of proposals in mind as we attempt to fairly manage the space environment in a responsible way for the future.

[100] International Telecommunications Union.
[101] International Civil Aviation Organization.
[102] Committee on the Peaceful Uses of Outer Space.

5.1 Data transfer issues

Wolfgang Rathgeber

Space activities are often more expensive than a single space actor can afford. Besides, they are inherently global by nature, and they are increasingly being recognised as a tool of diplomacy.[103] Accordingly, in the space sector there have been cooperation schemes at the international level or between partners at other levels right from the beginning. These cooperation schemes can include all phases of project execution: They might range from concerted system definition, design and development to joint operation of missions; from shared funding to a common utilisation of data gained by space activities. Data often represent the major objective of such activities, so their handling will be discussed in more detail here.

5.1.1. Terms and definitions

Two major modes can be distinguished: data sharing and data exchange. Data sharing is monodirectional; one party transfers some or all of its data to other parties without receiving data in return. Of course, data are a special commodity, because they can be copied. Consequently, they are not lost if they are passed on. Data exchange, in turn, is bidirectional. Two or more parties transfer some or all of their data to each other, ideally in a complementary way. Both data sharing and exchange will be considered in the following.

Data transfer should not just serve practical purposes, but must be based on more general principles like fairness as well. The concept of fairness has been introduced before. Fairness represents the ethical dimension of data transfer. By organising data sharing and exchange in a fair way, a responsible and sustainable use of space can be supported. Fair transfer regimes give an incentive to space actors to contribute to a global stock of data. They also enable space newcomers to access critical space-based data that they cannot generate on their own.

In this regard, the allocation basis deserves closer attention. Allocation can be based on equality, meaning that each partner gets the same share. It is questionable, though, if this can be considered as fair. Another possibility is basing the allocation on the needs of each partner. However, this raises the question how to measure the needs of donors and receivers – if there is an unbiased way of quantifying needs in the first place. Allocation can also be based on the amount of contribution – the more a partner offers, the higher the amount of data it receives. Again, this leads to the question of specifying the contribution in an objective way. Absolute scales might be used, but these will not reflect the available capabilities of each partner. The latter are better accounted for by using relative scales, with remaining ambiguities regarding the maximum capacities of each partner.

Data transfer occurs between different partners. The simplest case is given by cooperation of similar entities, as between countries at the international level or between companies at the national level. Such transfer regimes can usually be based on bilateral agreements. Transfer gets more complicated when unlike partners are involved and third-party interests or overarching principles have to be accounted for. In this case, trade-offs are necessary. Possibly contradicting interests can occur in cooperation between military and civilian entities, public and private entities as well as commercial and non-profit entities. This categorisation of trade-offs is not always clear-cut. Most of the trade-offs show up at the same time, as will be seen in the practical examples for common data utilisation that are discussed below.

5.1.2. The environment of data transfer

In any case, data sharing or exchange will have to respect various boundary conditions that exist along different degrees of freedom. Again, these categories are not completely separable. In fact, most of them appear simultaneously. One axis of orientation is the *organisational framework*. It comprises questions like:

- architecture: what is the functional set-up of the system, and where are the assets physically located;
- institutions: what entities are involved in planning, operation and oversight;
- data policy: what are the rules and procedures for handling and distributing information collected by sensors (raw data) and manipulated or refined in subsequent system stages (processed data) and what are the mechanisms for controlling and enforcing compliance with these rules and procedures; put shortly, who accesses what data under which conditions.

Another axis is constituted by the *technical framework*. This includes issues like:

- hardware: what kinds of sensors and other types of assets are involved?
- format: what physical representation of information is passed on – raw data (the notion of which is also subject to definition) or processed data?

- standardisation: is there a suitable uniform way of stating information to facilitate a smooth and efficient transfer?
- data flow: how can the data stock be administered and pooled, how is the transfer realized technically and what level of security and what kind of encryption is applied?

A third axis is given by the *political framework*, accounting for:

- cooperation: are there alliances to be served; or inversely, are there partners to avoid?
- economic interests: how can investments and resulting profits be protected?
- security: passing on sensitive information can endanger the well-being of individuals or States.

Last but not least, the *legal framework* is a decisive axis. In some regards, it is not yet fully adapted to the new situations arising from modern space applications. However, some efforts to update the legal regime are underway. Recent examples include the German satellite data security act that will be discussed below and the European Union INSPIRE initiative regulating the use of public geo-information service.[104] Generally, the legal framework is concerned with topics like:

- access: who is entitled to see the data?
- privacy: in how far are personal(ity) rights at stake and how can individuality be respected?
- (intellectual) property rights: can there be an obligation to pass on data that a partner has collected on its own and how can the ownership be protected?

5.1.3. Real-world examples

To give a flavour of real-world data transfer, some examples will be discussed here. The first one comes from the area of Space Situational Awareness (SSA).[105] SSA refers to knowing the location, motion, function and status of space objects as well as to realising the threats they might face. As such, Space Situational Awareness is an inherently global task. Some space actors like the U.S. and Russia run systems for Space Surveillance, which can be seen as a precursor stage of SSA. Europe currently does not dispose of SSA, although some systems that could be used for that purpose exist at the national level. Plans to set up a European SSA system are under way. For the moment, though, Europe gets the major share of its SSA data

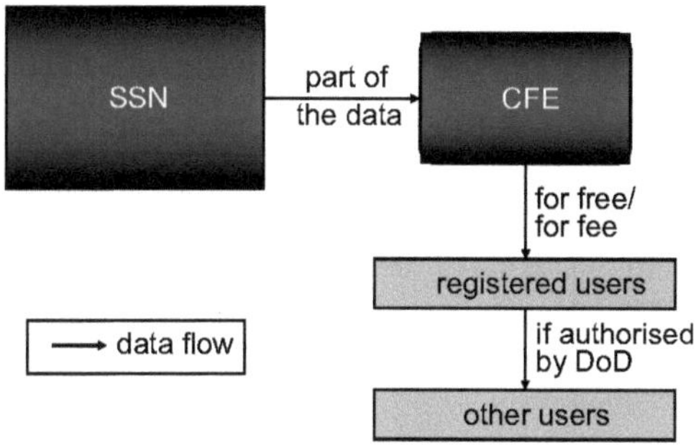

Fig. 1. *CFE data flow.*

from abroad, mainly from the U.S. The standard mechanism for that transfer is the Commercial and Foreign Entities (CFE) Programme.

Within this programme, the U.S. Air Force offers an unclassified subset of data from its Space Surveillance Network (SSN) via a website. Basic services are provided for free, while more advanced services have to be paid for. To gain access, users have to register stating personal or corporate information. By registering, users agree not to transfer data to third parties without permission by the U.S. Department of Defence. To gain this permission, they must specify the recipient, the further utilisation, the number of times they will share the data and the means of redistribution. Figure 1 shows the schematic flow of data for this example.

One could try to assess this data transfer regime in regard to fairness. Europe (with the exception of the United Kingdom and Norway) does not feed data into the U.S. Space Surveillance Network. Consequently, the CFE is an example of data sharing, not of data exchange. Applying the contribution basis to allocation, there is no moral obligation of the U.S. to share their data, especially given the fact that Europe is a space actor that could afford its own SSA system. However, beyond ethical categories it makes sense for the U.S. to give away part of their data to others in order to ensure safe operation of space assets for all space actors. It remains to be seen how the realisation of a European SSA system and the handling of the information gained thereby will affect the U.S.' willingness to exchange data.

A second example of data transfer is constituted by the German high-resolution Radar satellite system TerraSAR-X. It has been set up in a Public–Private–Partnership (PPP) between the German Aerospace Centre (DLR) and EADS Astrium. In return for contributing to the development costs of the satellite, Infoterra (a 100% subsidiary of EADS) holds the exclusive commercial utilisation rights. Regular customers are served through different distribution mechanisms. A special case is constituted by customers disposing of their own reception facilities. They can purchase a license and access the satellite directly.

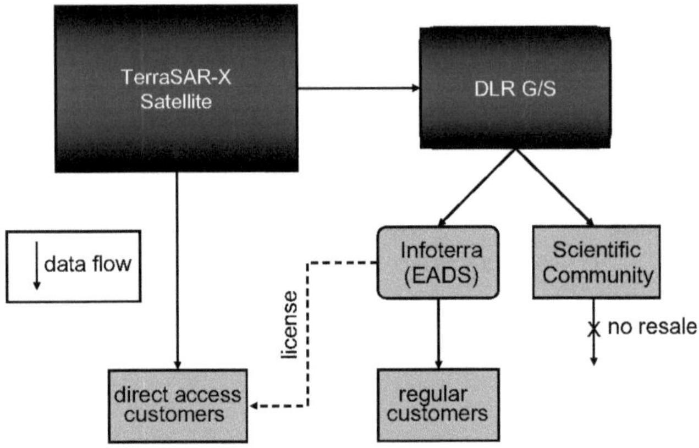

Fig. 2. *Data flow TerraSAR-X.*

DLR, representing the State, is the owner of the data. It is in charge of distributing the data to the scientific community both inside and outside DLR. The scientific community is not allowed to redistribute or to resell the data to third parties in order to protect the business case of Astrium. Figure 2 shows the schematic data flow for the system. TerraSAR-X will be complemented by a second satellite, TanDEM-X, that will also be realised in the framework of a PPP model.

Customer requests for TerraSAR-X imagery may be denied following the provisions of the German Satellite Data Security Act (Satellitendatensicherheitsgesetz). This act was passed by the German Parliament in view of the new generation of high-resolution imagery satellite systems.[106] Its aim is to ensure that the security interests of Germany and her allies are not endangered. The reference level for checking a possible danger is constituted by data that is freely available on the commercial market. The law incorporates a semi-automatic parametric approach: For each request, a numerical value is calculated. This value takes into account the requesting person or institution, the scene location, the processing parameters and the timeliness of the image. If the numerical value is below a certain threshold, the request is processed normally. If the numerical value exceeds the threshold, the request has to be authorised by the public institution in charge. This procedure allows for public control of sensitive requests while not interfering with the major part of the business. New and altered circumstances can be accounted for by adapting parameters and the threshold.

Again, it is worthwhile evaluating the fairness of this data transfer regime. The approach allocates data based on both contribution and need. It respects different interests – the economic one of Infoterra and the security-related one of the State, with the two possibly contradicting in some cases. Whether these interests are fairly balanced is a numerical question; it concerns the parameters defining a sensitive request and influencing the final authorisation as well as the share of budget that Infoterra had to contribute to gain the commercial utilisation rights.

The third example is not an existing one. It is an architecture that has been suggested for a future worldwide SSA system,[107] but it can also be applied to other space systems that are geographically dislocated or functionally extended. As mentioned before, SSA is inherently global. Surveillance of the geostationary ring, for example, is possible only if regions around the Earth provide their terrestrial observation capacities. Consequently, regional SSA systems will have to be merged to form an all-encompassing SSA network at some point in the future. This refers not only to systems as they currently exist in the U.S. or in Russia or as they are planned in Europe. Potential other future systems in Asia or Africa system will have to be integrated as well.

The regional systems can be linked together in a sub-centralised way, as shown in Figure 3. To this end, regional nodes can be created. Nodes in this context are understood as a more neutral notion than hubs, which are often associated with leadership or control. The nodes can be virtual or physically existing centres. They can serve as communication ports only or they can be charged with system guidance. In the case of Europe, the node can link national as well as European assets. The sketch is actually a top view of a three-dimensional data transfer regime, because the regional nodes can incorporate a multi-level approach: Within the regions, they can exchange data according to the regional data policy, and they can pass on a subset of the regionally collected data for inter-regional data transfer. At a higher level, the regional nodes themselves can be linked in a distributed or centralized manner.

Such a set-up facilitates international cooperation by allowing for implementing different data policies at the same time. This is important, because current or

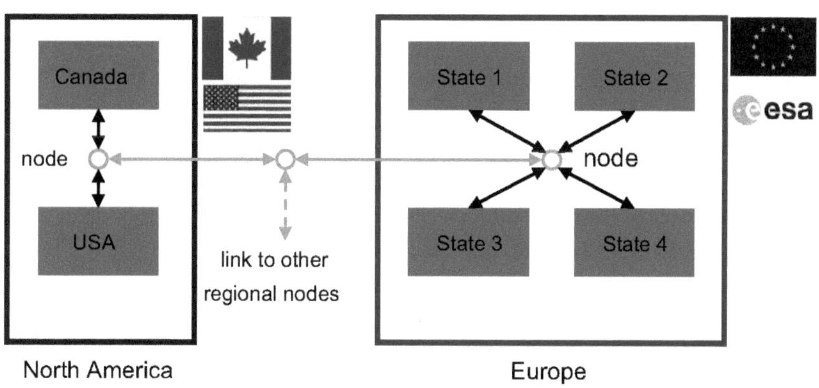

Fig. 3. *Potential architecture of global SSA system.*

imminent systems will not change their data policies fundamentally to serve a global regime. Rather, existing data policies have to be accounted for, for example by integrating them into an overarching umbrella data policy. While the proposed set-up is not fair by itself, it features favourable conditions for a give-and-take system serving the needs of everybody and creating an incentive to feed data into the system.

5.1.4. Assessment and way ahead

Given the examples above, a general assessment is in order. Although some efforts have been made and data transfer regimes are being recognised as being essential for global well-being, the current flow for space-based data is not sufficient. Too much information that should be accessible to relevant users anywhere in the world is gained and stored by different actors independently from each other, using separate systems. The situation is comparable to the Internet before efficient search engines like Google entered the picture. Space actors might not be aware that data they need are already available.

In addition, there are various obstacles that decrease willingness to pass on data and that hinder a free or at least a more abundant flow of information. Typically, reluctance to share or exchange data results from national concerns that are often associated with security issues. Within the European Union, security is within the realm of the Member States, and this is not likely to change any time soon. Classical security matters are handled by military and related entities, which feature a need-to-know culture. Naturally, this does not support generous data provision.

The occurrence of potentially contradicting goals is another impediment to data transfer. The balancing effort sometimes results in an overly restrictive handling of data. Moreover, transfer regimes cannot unfold efficiently when the legal background is not sufficiently clear. As was laid down, some new legal acts are being implemented, but the direction of development is not always uniform. There is a directive of the European Union stating that there is no obligation of public authorities to allow the re-use of their documents or data and that they are entitled to charge fees in order to facilitate a "reasonable return on investment".[108] This has the potential to add an economic perspective to public action, further complicating the structure of appearing interests.

A final point to be made for the assessment is that data for some domains like space weather, including solar activities, are simply not sufficient. This is regrettable, because an increased knowledge about space weather and its respective effects would be advantageous to all space actors. There are a limited number of missions to gain more information about these critical issues, but some of them like the European SOHO are about to expire with no immediate replacement foreseen. The scarceness of data hinders the set-up of an efficient data transfer system.

With this assessment, the next step is to ask what the future needs to bring and what steps have to be taken to achieve a more favourable situation. An important measure is to enhance the mechanisms for data sharing and exchange. This regards various aspects. First of all, an increased awareness of existing and available data is needed. This means that future data transfer regimes need to feature information about what kind of data are at disposal. This has to be supported by flexible and open data distribution mechanisms that are based on the willingness to pass on information. Broad availability of data will also foster a capable and competitive service industry. Finally, the sheer amount of data to be shared or to be exchanged has to be increased for certain domains like space weather.

Furthermore, an intensified international cooperation will bring about data transfer by itself. Such cooperation not only facilitates projects that single actors could not run on their own. It also expresses the concept of space as a common good of mankind. All nations should use it and the data resulting from its utilisation together. To achieve this goal, suitable mechanisms need to be employed, including financial ones. Developing countries and space newcomers should be supported not just as an act of charity, but as a long-term measure that will lead to new space systems worldwide, serving the well-being of the global community by adding novel and missing information.

Global systems for space-based data should be planned and set up in a truly coherent and strategic way right from the beginning. A coordinated approach will ensure that the single elements do not just avoid negative interference, but rather complement each other to achieve comprehensive coverage. Different data gathering tasks should be assigned to different actors based on their capacities and needs. If regional systems are to be combined to form a global system, unnecessary redundancy should be avoided. Duplication of core functions, though, will occur naturally as a consequence of national or regional strives for independence, and it should be used to avoid single-point failures.

Data sharing or exchange systems need to entail all kinds of possible present and future users. Potentially contradicting interests will have to be identified as early as possible in order to develop mechanisms for reconciling them. Military and other security-related entities will have to be involved in global data transfer regimes as well. In their case, some affinity for a need-to-share culture should be promoted, based on the insight that passing on data has the potential to increase the security of all partners involved.

The legal regimes need to be developed further to support global transfer of space-based data. This process could be flanked by formulating criteria for fairness that need to be met when sharing and exchanging information. As a critical enabler, data policies enshrine and codify the overarching principles of data transfer. In doing so, they mirror and enforce the underlying values. At the same time, they have to balance potentially contradicting interests, to respect the needs of different user communities and to settle questions of guidance, control and oversight. For this reason, they represent a core element of any data sharing or exchange regime. This has been and continues to be acknowledged in many space applications like the upcoming European SSA system. Efforts to ensure a fair way of data sharing will have to concentrate on setting up suitable data policies in the first place.

Data on space weather, especially information on solar activities, could be used as an important test case of a global transfer regime for space-based data. They represent a good opportunity to test policies and mechanisms of sharing and exchanging information. As in any area, the more the data are available, the better the knowledge. At the same time, there should be no reason to hold back information on space weather due to security concerns; so feeding all available data into the system would create an immediate win–win-situation. If the data transfer regime works well in this domain, the worldwide space community can move on to setting up other data transfer regimes in more sophisticated issue areas.

[103] Peter, Nicolas. "The EU's Emergent Space Diplomacy." Space Policy 23.2 (2007): 97–107.
[104] Smith, Lesley J. and Catherine Doldirina. "The EU INSPIRE Directive: A Suitable Mechanism to make Spatial Data more Available?" Presentation. 58th International Astronautical Congress. International Astronautical Federation, Hyderabad, India. September 2007.
[105] Rathgeber, Wolfgang. "Europe's Way to Space Situational Awareness". ESPI Report 10. Vienna, Austria. January 2010.
[106] Gerhard, Michael. "Security Interests in Earth Remote Sensing." DLR Countdown 5. Bonn, Germany. February 2008.
[107] Rathgeber, Wolfgang. "Europe's Way to Space Situational Awareness". ESPI Report 10. Vienna, Austria. January 2010.
[108] Council of the European Union and European Parliament. Directive Establishing an Infrastructure for Spatial Information in the European Community (INSPIRE). Doc. 2007/2/EC of 14 March 2007. Brussels: European Union.

5.2 Cooperation within the UN system

Luboš Perek

5.2.1. Introduction

Substantive cooperation in outer space matters has been in existence within the United Nations system of organisations since the 1970s. A large number of seminars has dealt with efficient use of applications of space technologies for the benefit of developing countries. Early this year the 28th session of the UN Inter-Agency Meeting on Space Activities, held in Geneva, reported on cooperative activities and plans.[109]

There is, however, an important field where stronger cooperation is desirable. It is the management of outer space. For instance, there is the problem of coordinating air-traffic with the flight of space objects through airspace. At present, no internationally agreed rules or practices exist. With the expected growth of space tourism it would be useful to start discussions of problems that might – and probably will – arise.

Another example is the relationship between space objects and space communication networks. Space objects are closely linked with space communications; yet space objects are registered by one organisation while space communication networks are in the hands of another organisation. The two bodies never compare the physical presence of space objects with the anticipated positions of space radio stations.

5.2.2. Space traffic through airspace

A year ago, the International Civil Aviation Organization, ICAO, published an important paper calling for extending the ICAO mandate, which currently applies only to airspace, to include outer space including the geostationary orbit.[110]

ICAO has unique experience in the following fields:

- airspace safety oversight;
- accident analysis and prevention;
- safety certification practices;
- coordination of flights in air.

Mutatis mutandis, these elements could be applied to space objects passing through airspace. The most striking gap is the present absence of certificates for space objects. In air-traffic and even in road-traffic, technical certificates are obligatory and on the sea "every sailor is entitled to a seaworthy ship". The most sophisticated vehicles are those used in space travel, yet they are not obliged to carry internationally recognised certificates. With the growing participation of private commercial companies in space activities it seems logical to institute obligatory global practices and standards.

The experience of ICAO could solve the old problem of formally adopting an upper limit of airspace coinciding with the lower limit of outer space. Indeed, in Section 1.2 on page 2, the ICAO White Paper states that the adoption of a limit is essential for regulatory and liability reasons for suborbital systems operating near the natural limit of airspace. The UN Committee on the Peaceful Uses of Outer Space, COPUOS, has discussed the problem of delimitation of outer space for over 30 years. Its Scientific and Technical Subcommittee ended its discussions in 1967 because "science cannot determine an accurate limit". This was rather a pretext than a reason. Many physical parameters of the atmosphere change continuously with altitude. Scientific and technical factors can indicate the maximum altitude an airplane using aerodynamic reaction can reach. That is the von Karman limit at about 83 km. On the other hand, satellites of current construction can complete one orbit around the Earth down to the altitude of about 100 km. The legal limit thus could be agreed upon anywhere between the above values. Its exact value is required for legal reasons in order to know whether at a given moment space law or air law applies.

The Legal Subcommittee has continued its discussions of the delimitation of outer space until the present time. It has discussed all possible options but unfortunately without achieving consensus.

As regards outer space up to the geostationary orbit at an altitude of about 35,900 km, the ICAO paper does not elaborate on principles of outer space management. Physical conditions in outer space are entirely different from conditions in airspace. For example, an object in airspace tends to fall to the ground whereas in outer space it tends to continue in its trajectory. Braking slows down an object in airspace but it accelerates an object in outer space. Economy of efforts, as well as tradition, supports the principle of separate management of airspace and of outer space.

Responsibilities for outer space are in the hands of COPUOS and the International Telecommunication Union, ITU. COPUOS was established by the UN General Assembly in 1959 as a permanent committee. Its mandate is very wide and includes the study of legal problems. Five international treaties and a number of recommendations have been elaborated by COPUOS. The UN

General Assembly can also be expected to support the future involvement of COPUOS in outer space matters.

Radio telecommunications between radio stations on satellites and radio stations on the ground are managed by the ITU, a specialised agency of the UN. The ITU became engaged in radio communication between ground radio stations as early as 1927. The role of the ITU in space matters is likely to continue also in the future.

5.2.2.1. Conclusion

The role of ICAO should be seriously considered for airspace, up to its border with outer space, where its expertise is beyond doubt.

5.2.3. Relation between satellites and space communication networks

5.2.3.1. Satellites

The need to possess information on the launching of objects into outer space ("the need to know what is up there") was the first requirement for international cooperation in space matters in history. It was included in UN General Assembly resolution 1721(XVI) in 1961. It culminated in the adoption of the Registration Convention in 1974. That Convention was elaborated in the climate of the Cold War. An agreement was possible only on rather general matters. The Convention does not fix exactly when the announcement of the launch of an object into space has to be submitted. It uses vague wording "as soon as practicable" which allows loose interpretation. Some launching states comply with the spirit of the Convention and register within a few months; others register after one or more years, if at all. The following table gives an idea of the recent practice of registrations:

Year of launch	2004	2005	2006	2007
Total number of objects launched	72	71	95	111
Percentage of objects not registered (%)				
by August 2007	17	15	38	–
by October 2008	15	7	23	39

Registrations have been received at the UN Office of Outer Space Affairs as long as three years after launch, but the register is still incomplete. In December 2007, the UN General Assembly adopted a resolution[111] on enhancing the practice of States and organisations in registering space objects.[112] The resolution recommends, *inter alia*, the use of the COSPAR international designator as a reliable aid to identification, the adding of appropriate information on the location in orbit of a geostationary satellite, its transfer into a disposal orbit, any change of operational status, and several additional improvements. It is still too early to note the impact of the resolution on practices of registration.

Launching states and international organisations should use the opportunity to make the UN register, which is set up in accordance with the Registration Convention, the most authoritative source of information on space objects. At present, there are other faster and more reliable sources of information. For example, the OOSA maintains an Online Index of Space Objects reporting data on registrations as well as data on non-registered objects.[113] The Index is, as a rule, updated frequently. Other sources are the monthly Spacewarn Bulletin[114] and the Encyclopedia of Satellites and Probes.[115]

The most important and most frequently used satellite orbit is the geostationary orbit with its well-known advantage of needing only fixed ground antennas for establishing radio connection. To keep a satellite in this orbit, natural forces, in particular the attraction by the Moon as well as by the irregularities in the shape of the Earth, have to be counteracted by periodical bursts of small rockets. This practice is called "station keeping".

Orbital elements of geostationary satellites, based on observations by a worldwide network of telescopes and radars known as the NASA Two-Line-Elements, are being used for computing actual positions of GEO satellites. These are published annually by the European Space Operations Centre in Darmstadt.[116] Individual positions, based on the same observations, appear also in the Encyclopedia of Satellites and Probes (see footnote 115). The actual number of active satellites in the geostationary orbit changes frequently because new satellites are launched and old satellites are deactivated. An approximate number in mid-2008 was 370.

5.2.3.2. Space communication networks

Space communications are within the mandate of the ITU. Its constitution states in Article 12 that the mandate is to ensure rational, equitable, efficient and economical use of the radio frequency spectrum by all radio communication services – including those using the geostationary satellite orbit or other satellite orbits – and to carry out studies on radio communication matters. Preventing harmful interference is a strong enough reason for States (called "administrations" in the ITU documents) to submit a request for international recognition and protection to the ITU and to abide by the coordination process with other users of the spectrum. The ITU keeps records of the coordination process. The main results are currently published in the quarterly Space Network List and the Space Network List Online.[117] There are three main stages of processing the requests: Class A denotes the receipt of an announcement by the ITU, Class C refers to coordination with existing networks and Class N is used for networks after successful conclusion of the coordinating process. The symbol N stands for "notified" in the Master International Frequency Register. Class N networks enjoy international recognition and protection from harmful interference. The work involved in the coordinating process is enormous. The number of entries in recent issues of the Space Network List has exceeded 5700, because some networks appear in more than one class. The number of separate networks is about one half of the total number, with only 800 networks in Class N. Adding some 80 networks of planned broadcasting services, the number of N networks is about 880.

The ITU has introduced a special terminology, slightly different from current use. For instance, a "space station" in current use means a huge international station with a permanent crew, while for the ITU it means a "radio space station", which is a radio receiver or transmitter located on a satellite. The term "space station" has been used in ITU Reports to COPUOS.[118] In a recent edition of the SNL, another term, "Sat_name" or, in full, "Satellite Name" was used for the heading of a column identifying space networks. An explanation of the term in a preface does not remove a casual reader's impression that the column refers to satellites. In fact, the distinction between a satellite on one hand, and a radio space station or space network on the other, is essential. A satellite is a vehicle in orbit carrying on board many component parts, among them one or more radio receivers or transmitters. A satellite can carry more than one "radio station". Vice versa, one radio station can be operated on more than one satellite and its function can be transferred, e.g., from an old satellite to its new successor at the same orbital position.

5.2.3.3. Comparative table[119]

The real situation in the geostationary orbit is best seen by comparing radio space stations of space networks with satellites present and active at the same orbital position. The main results are the following:

- At a given nominal position an active satellite would be present and there would be one or more space networks of Class N transmitting radio communications from that satellite. Unless additional information is available, it cannot be determined which of the networks are operated by that satellite. In some – but not all – cases, such additional information can be found on the Internet.

- If more satellites are present at a nominal orbital position, additional information is needed to determine which satellite is operating what space network.

There are two possible abnormal cases:

- If no satellite is present at the position in question, no transmission from that position is possible and the space network cannot be operated even if it is in Class N.
- If a satellite is present, but no Class N space networks appear in a relevant list, possible radio transmission has not been coordinated with existing space networks and there is a danger that harmful interference may occur. If there is no interference, transmissions are physically possible but this would not constitute the orderly conduct of space activities.

A comparative table of space networks and satellites has been prepared and published in a paper of the Scientific-Legal Roundtable on Paper Satellites on the

SPACE NETWORKS - NOTIFIED						SATELLITES IN GEO			
Nom. Long.	Adm	Ntwk org	Space Network Name	Frequency bands		COSPAR Int. Desig.	Satellite Name	Mean Long.	Status of orbit
0.00 E	F	ESA	METEOSAT	3-33	•	2005-049B	MSG 2 Meteosat 9, i=0.93	0.05 E	C2
	F	ESA	MSG	17-35	•		MSG 2		
	USA		USCID-A1	66-79					
1.00 E	RUS		GALS-15	49, 50					
	RUS		STATSIONAR-22	43-47					
.	RUS		VOLNA-21	9-13					
2.90 E	CTI	RAS	RASCOM-1F	43-55	•	2007-063A	Rascom-QAF1	2.88 E	PL
	CTI	RAS	RASCOM-2F	43-55			Regional services in Africa		
3.00 E	F		SYRACUSE-3F	33-82	•		Telecom 2C		
	F		TELECOM-2C	33-57	•	1995-067A	Telecom 2C, i=3.82	3.05 E	C2
	F		TELECOM-3C	43-48					
4.00 E	F	EUT	EUTELSAT 2-4E	33-57		1997-049A	Hot Bird 3 = Eurobird 4	4.02 E	C1
	F	EUT	EUTELSAT 3-4E	33-57			Regional services in Europe		
	F	EUT	EUTELSAT-KA-4E	76-78					
	F		SMO-GEO-1B (BSS)						
	F		TELECOM-4E	47					
	USA		MILSTAR-13	31-84					
	USA		USGAE-2	31-84					
4.80 E	S		SIRIUS-2 (and BSS)	33,57	•	2007-057A	Sirius 4	4.83 E	PL
	S		SIRIUS-4.8E-BSS			Services in North and East Europe, North Africa			
	S		SIRIUS-4.8E-BSS-2			1993-031A	Astra 1C, i=1.21	4.74 E	C2
5.00 E	S		SIRIUS-30B	52.55		1998-056B	Sirius 3	5.00 E	C1
	S		SIRIUS-30B-5E	43-55		Services in Scandinavia and Greenland			
	S	NOT	TELE-X (and BSS)	33-57					
	S		SIRIUS-5E-BSS						
	USA		USMB-5	35					
5.70 E	MLA		MEASAT-SA1	43-48					
6.00 E	G		SKYNET-4B	9-82					
	G		SKYNET-4K	9-35					

Fig. 4. *Comparative table sample page.*

DVD of the International Astronautical Congress held in Glasgow in 2008. The results found for a reference date of 30 June 2008 were surprising:

- A total of 165 notified space networks (19% of the total number of 880) at 83 different orbital positions had no satellite at their respective positions. This means that at least one out of every five notified space networks was not in operation.
- A total of 23 satellites had no notified space networks at their orbital positions. Some of these satellites were transmitting.

5.2.3.4. Conclusion

The spirit of the Registration Convention as well the spirit and letter of the relevant ITU documents are not respected by all launching agencies and satellite operators. For various reasons, exceptions to any rule have to be expected. However, the numbers found in simple comparisons of the ideal and the real state of affairs looks more likes the rule than the exception. The question is: Can COPUOS and the ITU cooperate to use the Comparative Table to assess the real use of the GEO and to turn the practice of bypassing existing rules into a new set of rules better corresponding to the existing situation and to the actual needs of users of the GEO?

[109] See United Nations General Assembly.Report of the United Nations Inter-Agency Meeting on Outer Space Activities on its twenty-eight session. UN Doc A/AC.105/909 of 23 Jan. 2008. Vienna: United Nations.

[110] International Association for the Advancement of Space Safety. "An ICAO for Space?" 29 May 2007. 18 June 2009. http://www.iaass.org/pdf/ICAO%20for%20Space%20-%20White%20Paper%20-%20draft%2029%20May%202007.pdf.

[111] United Nations General Assembly. Recommendations on enhancing the practice of States and international intergovernmental organisations in registering space objects. UN Resolution 62/101 of 17 Dec. 2007. New York: United Nations.

[112] For details of Resolution 62/101 see Schrogl, Kai-Uwe and Niklas Hedman. "The GA Resolution 62/101." Journal of Space Law 34.1 (2008): 141–161.

[113] UNOOSA website. 18 June 2009. www.unoosa.org/oosa/osoindex.html.

[114] "NASA Spacewarn Bulletin." NASA. 18 June 2009. nssdc.gsfc.nasa.gov/spacewarn.

[115] "Encyclopedia of Satellites and Probes." 18 June 2009. www.lib.cas.cz/space.40.

[116] Choc, R. and R. Jehn. "Classification of Geosynchronous Objects." February 2008. European Space Operations Centre. 4 Nov. 2009. http://lfvn.astronomer.ru/report/0000028/index.htm.
[117] "Space Network List." ITU. 18 June 2009. www.itu.int/ITU-R/space/snl.
[118] For example, in: International Telecommunication Union. Thirty-fourth Report on Telecommunication and the Peaceful Uses of Outer Space. Booklet No. 34. Geneva: ITU, 1995. Table of Geostationary Space Stations: 21–40.
[119] The Comparative Table lists 880 space networks and 372 satellites at 314 orbital positions. Because of its size it cannot be printed here. It will be sent on request addressed to perek@ig.cas.cz.

5.3 Current international space security initiatives

Ben Baseley-Walker

Heightened awareness of the importance of space security as a crucial building block for the long-term sustainability of outer space has led to a shift in the international community in its thinking on how to approach space activities. The upshot is the current proliferation of ideas and initiatives on how to improve security in the space environment. This article lays out some of the foundations that are being discussed in those initiatives and details some of the approaches that are being taken.

The current wave of international proposals can be broken down into two groups – "top-down" initiatives and "bottom-up" initiatives. Top-down initiatives are focused on creating structures built around international regulation or legal instruments. Bottom-up initiatives are typified by approaching space security topics from an applied technical perspective focusing on solving problems facing those working and operating in the space field. Past engagement on space security has been predominantly via a top-down approach; yet over the past few years the international community has seen an increasing focus on a more directed, problem-solving approach. This can be ascribed in part to the fact that space is now becoming ever more ubiquitous in so many sectors of human activity and, as more and more countries enter into the space world, the need to establish an effective methodology for space security for all spacefarers is becoming ever more pressing.

From the policy perspective, space security has become increasingly prominent. There are, however, several key considerations which affect space security but are not derived from the politics of space. First amongst these is the question of organisational proliferation. Over the last 20 years or so, there has been a move in the international community away from the creation of new international organisations to undertake management or regulation of emerging international issues. After the creation of many new organisations at the end of the Second World War, the costs of maintaining these organisations and their increasing complexity turned attitudes away from building brand new organisational structures.

As such, the international community of today no longer subscribes to a base position of automatically creating a new international organisation to deal with an international problem. This shift has been mirrored in the type of agreement that codifies international agreements on a particular topic. The traditional default of establishing a binding legal agreement, like a treaty, is a time-consuming and laborious process. A growing interest for voluntary and non-binding agreements has diversified the options that can be considered when assessing how we, as an international community, might like to proceed on managing a particular issue. In the space sector, this is seen most clearly in the United Nations Debris Mitigation Guidelines,[120] a non-binding voluntary agreement designed to limit the production of space debris. Whilst this agreement did not build a regime of binding standards for the international community at large, it has significantly influenced the activities of the major space faring States and has created an increased level of understanding of the debris issue globally. This has also been reflected in the norms and laws implemented in the domestic environments of some key space States on mitigation of production of space debris.

Last, the current international political climate impacts how we can possibly move forward on these issues. The professed desire for increased engagement with the international community of the new United States administration has the potential to move forward many issues, including the cause of space security. In that undertaking, it is important to take account of the increasing number of space players and those who benefit from space services. The key question as to how does one achieve a balance between the interests of those States already heavily invested in the space environment and emerging space States should always be kept in mind when analysing new space security initiatives.

5.3.1. Space security initiatives and international law

In understanding new proposals and how they fit into the international environment, it is beneficial to bear in mind some of the most basic principles of international law, including the differences between types of international law and what we can understand from choices that States make regarding the types of international agreements they choose to sign up to. For purposes of this article, the three major types of international law can be said to be treaty law, customary international law and soft law.

In short, treaties are legally binding agreements in which States agree to certain courses of action or interpretations of a specific question. Perhaps the most important concept when thinking about treaties as a possible method of managing international issues is that of "consent to be bound". International treaties are often misrepresented as a "rule" that is imposed on parties from above, much in the way that we think about domestic laws. This, however, is a misconception. A treaty is better characterised as the confluence of the interests of two or more States. At the basis of every international treaty, every State that signs up to that treaty indicates consent to be bound by the terms of that treaty as a representation that the treaty reflects the aims, intentions and policies of that State in the international environment. As we have seen, when those interests and positions change, we do at times see States withdrawing from such agreements such as the 2002 withdrawal of the United States from the 1972 Anti-Ballistic Missile Treaty.

The second form of international law is customary international law, which can be defined as "a general practice accepted as law".[121] Further, there are two elements that must be fulfilled in order to declare that a certain course of action is customary international law: (1) A specific course of action must be widely practised amongst States and (2) A belief held by these States that they are obligated to carry out such a course of action, or in legal terminology, *opinio juris*, a belief that something is law. An example of customary international law would be a prohibition of the use of force in international activities. It should be important to note that customary international law is considered binding on all members of the international community except in very specific circumstances.[122]

The last source of law for the purposes of this brief overview, and one that is currently very relevant as it is the chosen method for many current space security initiatives is soft law. The key element of soft law, often contrasted to the "hard law" of a treaty, is that it is non-binding. In the space security realm, a very good example is the UN Guidelines on Debris Mitigation which are voluntary but have been accepted by the General Assembly of the United Nations.[123] Soft law can be taken at a later point as contributory evidence that a particular course of action has become customary international law.

This leads us to why these major types of international law are relevant to an analysis of space security initiatives. First and foremost, it clarifies the concept of enforcement in international law. There is no global policeman ready to pounce when a treaty or customary international law provision is transgressed. That, in sharp contrast to the domestic legal context, is not the nature of the system. As such, keeping in mind that international legal instruments are based around international acceptance of, and consent to, particular principles, it allows us to analyse the value of current moves towards improving space security from an international perspective. In that process one should recollect that enforcement in international law is not a central pillar of the system and most often involves more unconventional "sticks".

5.3.2. Top-down initiatives

To highlight the nature of top-down initiatives and how they approach improving space security, here two different models are examined, the

Chinese–Russian proposal for a Prevention of an Arms Race in Outer Space (PAROS) Treaty and a model for an international organisation for space based on the International Civil Aviation Organization (ICAO). Whilst very different, both initiatives start from a principle of creating overarching international structures to achieve their stated goals of safer space.

5.3.2.1. The Chinese–Russian treaty proposal

The main elements of the Chinese–Russian treaty proposal are a ban on weapons placed in outer space, the creation of an international monitoring and enforcement agency, and a ban on threats or use of force in space. Referring back to the overview of international legal sources above, the Chinese–Russian effort proposes a binding legal regime that would require States to consent to be bound by the terms of the treaty.

There are some concerns with the treaty proposal. The first of these is focused on the way that the PAROS treaty has been worded. It bounces back and forth from being very vague as is mentioned in Article VII, "the State Parties, shall practice on a voluntary basis, unless otherwise agreed, confidence-building measures",[124] to over-specification of definitions such as the definition of a space weapon in Article 1,

> "*the term "weapons in outer space" means any device placed in outer space, based on any physical principle, specially produced or converted to eliminate, damage or disrupt normal function of objects in outer space, on the Earth or in its air, as well as to eliminate population, components of the biosphere critical to human existence or inflict damage to them.*"[125]

A concern is that the treaty is much too general yet at the same time very prescriptive on certain things. Faced with the realities of the advance of space weaponisation such a broad yet specific treaty could in theory fail to limit weaponising activities. One of the perennial debates in international space policy is how to define a space weapon. Fundamentally, nearly EVERY object with manoeuvring capability could potentially be used as a space weapon.

Crucially, the Chinese-Russian proposed treaty does not cover ground-based anti-satellite weapons. The definition in the treaty outlined above requires that a "space weapon" must orbit the Earth once or follow a section of such an orbit before leaving orbit.[126] This seems to specifically exclude ground-to-space kinetic anti-satellite weapons which are the type of anti-satellite weapons that have been developed by Russia in the past and are currently being developed by China.

Lastly, referring back to the oft-voiced concern in the international community of organisational proliferation, the treaty creates a new international organisation with no clearly defined purpose and mandate. Overall, it seems that the Chinese-Russian proposal takes many existing commitments and repeats them, for example the prohibition on the use of force,[127] yet also lacks clear definition on new concepts.

5.3.2.2. An "ICAO for space"

In sharp contrast, the proposal of the International Association for the Advancement of Space Safety (IAASS), "An ICAO for space" is very specific. Its key provisions can be summarised as the creation of an ICAO-like[128] organisation for space either as a sub-branch of ICAO or an independent institution designed to regulate space activities in the international realm. ICAO currently oversees traffic management and coordination for civil aviation. The proposed IAASS organisation would aim to ensure that citizens of all nations are equally protected from "unreasonable levels" of risk from overflight by missiles, launch vehicles and returning spacecraft. It would also endeavour to ensure that any spacecraft would be developed, built and operated according to uniform minimum safety standards which reflect the status of knowledge and accumulated experience. It would strive to prevent the risk of collision or interference during transit in the airspace and on-orbit operations and it would aim to ensure the protection of the ground, air and on-orbit environment from chemical, radioactive and debris contamination.

Whilst this sounds positive in principle, it would seem to involve a much greater degree of integration of systems and approaches than seem to be

Fig. 5. *ICAO logo (source: ICAO).*

currently politically feasible in the international community. The likelihood of implementing such a wide-ranging system given the national security and political interests of many of the States that would need to be involved is questionable. The United States, for one, would seem loathe to submit itself to such a regime, given its views on freedom of action in space. Last, coming back to two of the key themes that have been discussed here, the ICAO proposal again raises the questions of organisational proliferation and an inaccurate conceptualisation of the reach and validity of enforcement mechanisms for improving space security in the near to medium term (Figure 5).

5.3.2.3. Top-down initiatives: fair and responsible?

In summary, how can we assess the value of top-down initiatives directed at an amelioration of security in space? Since outer space is one of the world's most globalised environments, the process is fully international and all States can engage in such a process should they so wish. This is a significant advantage of the top-down model. However, a key question to ask is whether emerging space States and States that have a desire to take advantage of space resources in the future have the capacity in Geneva[129] to make effective contributions to such a document. Furthermore, it would seem that such initiatives are focused on the agendas of the major players and the evidence of key *lacunae,* especially in the PAROS treaty proposal, is notable. Last, the concerns of potential human capacity constraints in the implementation of the structures laid out in concepts in the vein of the *ICAO for space* model may provide a barrier to implementation for new entrants to the space arena and undermine the effectiveness of such an approach.

5.3.3. Bottom-up initiatives

In contrast to the overarching schemes for space security regulation, over the last year or so, we have seen a different approach gaining traction – that of the bottom-up initiative. The concept of bottom-up describes those efforts which focus more on technical analyses of and proposed solutions to specific problems. These initiatives are characterised by an approach that often does not see a comprehensive international treaty as the end goal. Here, three initiatives have been selected derived from three very different parties: the "Paris working group" principles developed by a cross-section of international stakeholders, the draft proposal for an international code of conduct developed by the European Union (EU), and the data-sharing initiative developed by the major commercial satellite operators. The selection of these three aims to give an overview of the cross-the-board interest in applied regulation as opposed to traditional international endeavours.

5.3.3.1. The working group of experts on the long-term sustainability of space activities

The Working Group of Experts on the long-term sustainability of space activities principles can be conceived of as a set of guidelines on operational activities in space with an eye towards ensuring long-term sustainability of space for human activity. The Working Group of Experts itself was an informal working group set up prior to the Scientific and Technical Sub-Committee (STSC) of the United Nations Committee on the Peaceful Uses of the Outer Space (COPUOS) session in 2008. The group is currently in the process of developing an extensive document of recommendations from technical experts. The genesis of this process owes much to the success of the Inter-Agency Debris Coordination committee (IADC). The IADC's development of the Space Debris Mitigation Guidelines over the course of more than a decade was a success. The IADC used a fundamentally dual-track process with one foot inside the UN and one outside, and its success has acted as an impetus for support for the dual-track approach. As such, the Working Group of Experts adopted this approach. Comprised of a group of key international actors from government and the space industry, the group is focused on a pragmatic approach to key issues facing space actors. These key themes include the question of space debris, including national implementation of the UN Space Debris Guidelines, increased orbital crowding, spectrum management and space weather.

The Working Group of Experts is an interesting model. As with most of the bottom-up initiatives, it focuses on voluntary "best practices" rather than creating a regulatory system. It is of some concern, though, how exactly such a set of best practices or principles would relate to other such initiatives currently being discussed in the international community, most notably, the EU proposal for an international code of conduct. Is there a risk that we are developing too many types of operational guidelines and that a sector that until recently had very little clarity will soon become over-complicated by a plethora of "how-to" efforts?

In the case of the Working Group of Experts another concern is its dual-track nature. Whilst the IADC developed its technical guidelines outside of the UN and then wrapped them into the UN system in their search for international acceptance, the Working Group of Experts has been more engaged in the UN system from the outset. This is potentially of concern as, given the nature of the COPUOS as a consensus body, there is a substantial risk that politics may overtake practical substance.

From an international law perspective, such a process is not legally binding on parties that sign up to it. If the principles that the group eventually develops are proposed in stand-alone manner or if they are incorporated into the UN, they would remain soft law.

One of the other questions with the Working Group of Experts is whether or not it truly represents the equities of all space players. This state of affairs highlights one of the key differences between the top-down and the bottom-up approaches. The Chinese-Russian treaty proposal incorporates nearly all international stakeholders; bottom-up initiatives tend to be derived from smaller groups with aligned interests. The difficulty then is to translate the agreements and common frameworks developed in these aligned sub-groups and gain international engagement and acceptance of such principles. To reiterate, space, as one of the world's most globalised environments, requires engagement by all stakeholders, as the activities of a sole actor can have a significant effect on the ability of the community as a whole to continue to use space resources.

5.3.3.2. The European Union draft proposal for an international code of conduct

The EU, in regard to its proposal for an international code of conduct, has grappled with this very question – how to expand a European-driven initiative to involve all the necessary international players. The proposal itself involves a commitment to make progress towards adherence and full implementation of current treaties and norms relating to space security, a commitment to prevent space from becoming an area of conflict, recognition that space is essential to national security and strategic stability and the need to develop conflict resolution processes which recognise the inherent right of self-defence. Specifically, the proposal deals with operational questions such as collision avoidance, preventing deliberate explosions, improved space traffic management, data exchange and confidence-building, notification of action and a push for more stringent debris mitigation by States.

The proposal is sound in many ways and is laudable in that it is evidence of an increased European interest in space security, especially when coupled with

European moves to integrate their space situational awareness (SSA) capability. However, concerns with the European proposal have come to light most obviously in the process of taking the concept to the broader international community. There has been significant discussion as to the forum in which the European code of conduct would be presented to the international community, namely, whether it should be at the United Nations Conference on Disarmament (CD) or COPUOS. As both are consensus-based, there has been much concern as to the success of the European proposal because of the competing political pressures of the State members. At the time of writing, the current suggestion for the internationalisation of the proposal is that the code of conduct would be a free-standing agreement that would not be associated with any international body.

The parallel that is often made is that of the Hague Code of Conduct on Ballistic Missile Proliferation (HCOC). Under the HCOC model, there are no limitations on who can be a subscribing state; additionally, States that do become signatories make non-binding commitments to follow the tenets of the code. HCOC is currently subscribed by 130 States. Whilst this could in fact end up being a successful model, it does clearly highlight one of the key problems with bottom-up initiatives – one may end up with only a select few supporting such an initiative. In the space environment, this could prove counterproductive. In the case of space security, it is important to bear in mind that the boundaries of what constitutes acceptable interaction are much less clear than in the field of ballistic missile proliferation. What remains to be seen is whether the European proposal, if it indeed propagated in the HCOC mould, can garner enough political agreement in the international community, and therefore signatories, that the code gains sufficient legitimacy (Figure 6).

Fig. 6. *Missile launch (source: Telegraph Media Group Limited).*

5.3.3.3. Satellite operator coordination initiative

The last initiative discussed here is perhaps the one that breaks most from the traditional models of top-down initiatives in international relations. The satellite operator coordination initiative is a consortium of several satellite operators which aims to standardise orbital prediction models and reporting requirements to facilitate data exchange. The initiative has developed a "babelfish" programme for "translating" information from one company's data model to another. Currently, a prototype system for reporting data to a "neutral" third party, the Center for Space Standards and Innovation, has been developed. This third party aggregates the data from all the participants and performs critical analysis functions. One of the key elements of the initiative are the provision of technical support for close approach mitigation and the group is working on providing automatic close approach or collision warning including with space debris (Figure 7).

The satellite operator coordination initiative is the model, certainly amongst all those examined in this chapter, which is closest to the on-the-ground activities of day-to-day space operations. This particular form of coordination was derived out of business necessity as opposed to political regulation. In bottom line terms, it is in the interests of the commercial satellite operations to improve data access and space security in order to reduce their risk of losing space assets to collisions and thus reduce their risk of reduced profitability.

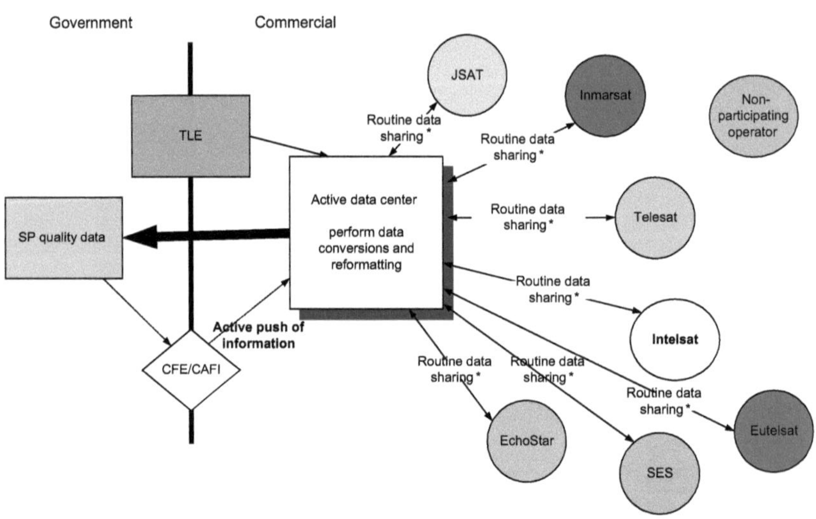

Fig. 7. *Satellite operators coordination initiative (source: Richard DalBello, Intelsat General).*

Whilst this initiative is focused on a specific issue, as a model of developing structures that deal with a key problem facing space actors, it is a clear step in the right direction.

There are some concerns with the initiative. Whilst this is an attempt to lower barriers to cooperation within the industry, there is a question how far such a model can be expanded out to State actors as opposed to those in the civil and commercial sectors. Given concerns of national security and the political relationships among certain key space players, can this system really provide an effective solution to the question of limited space situational awareness for many space actors? Does the propagation of such an initiative amongst a select group of actors decrease the drive towards finding a system that can involve all major space players? At its base, there are also questions as to the purposes of the satellite coordination initiative. The initiative acted as a fulcrum to reengage the US military in supporting the needs of the international commercial satellite operators and to what extent this coordination initiative was simply a timely political tool as opposed to an effective long-term measure remains to be seen.

5.3.3.4. Bottom-up initiatives – fair and responsible?

In summary, the recent focus bottom-up concept of moving towards international engagement on space security issues seems to be a development that is beneficial to the space security cause. Basing a working approach on key issues from positions of technical consensus is a sound model. From the point of view of international law, building non-binding international agreements expressing the confluence of international opinion would seem to lay the basis for principles that are likely to be adhered to in the long-term. There are, however, legitimacy issues and the key question of how one moves to incorporate all space actors in the bottom-up approach. In terms of whether or not the models discussed here are fair and responsible, the bottom line seems to be, who is involved in these initiatives and are they sufficiently internationalised? One of the successes of the "Paris working group" is that it has sought to bring in more than just State representatives. Conversely, one of the concerns with the satellite operators' coordination initiative is whether access would trickle down to all those who need it.

5.3.4. Concluding thoughts: the way forward

In the current international space environment, with the ever-increasing numbers of emerging space States, building international policy and potential agreements on the basis of sound technical foundations is strongly advisable. For the time being, a holistic top-down approach is unlikely to have success for two key reasons: a resistance to organisational proliferation and a general lack of commitment to space security at the domestic policy level. Likewise, a bottom-up approach also has a questionable level of success as on its own, it likely to lack international buy-in from key partners. This can be seen with the fact that whilst the UN Space Mitigation Guidelines were by the UN, the success of the implementation of the guidelines still hangs in the balance.

What can be suggested as potential options for ways forward? Given the concerns that exist with both top-down and bottom-up approaches, it would seem advisable to examine a hybrid model of engagement. An example would be specific agreements on key technical concerns such as the use of debris-causing kinetic energy anti-satellite weapons. What is clear from the current space security situation is that there is a demonstrable need for the technical community to engage with international policymakers. Currently, many of these initiatives are discussed in the international discussion fora, yet there is a dearth of diplomats who are well informed on this issue. Such an information campaign lays solid groundwork to moderate the kneejerk reaction that will inevitably occur after the next high-profile on-orbit destructive event. Such a campaign will augment effective discussion and action on defining the parameters of how entities interact in space to ensure its sustainability and long-lasting utility.

The central challenge that faces the international community will be bringing together all of the initiatives identified here into a network of effective agreements that regulates human interaction in space without hindering the entry of new players into the space arena. Whether this is formed from a multilateral process or a regional/bilateral network of commitments remains to be seen. As a final thought, it is worth stressing again one of the key underlying assumptions of this overview: it is imperative that all space actors are engaged and invested in space security initiatives. This is a fundamental requirement necessary in the building of a fair, responsible and sustainable methodology to continue to allow humanity to maximise their long-term use of the tremendous benefits that space has to offer.

---

[120] United Nations Committee on the Peaceful Uses of Outer Space. Space Debris Mitigation Guidelines. UN Doc. A/62/20 of 10 Jan. 2008. Vienna: United Nations. Adopted by: United Nations General Assembly. Resolution on International Cooperation in the Peaceful Uses of Outer Space. UN Doc. A/62/217 of 1 Feb. 2008. New York: United Nations.

[121] Article 38 (1) (b), Statute of the International Court of Justice.

[122] The persistent objector principle in which a State has consistently declared that it does not consider a certain provision to be customary international law and as such does not see itself bound by it. See Harris, DJ. Cases and Materials in International Law. London: Sweet & Maxwell, 1998: 42.

[123] United Nations Committee on the Peaceful Uses of Outer Space. Space Debris Mitigation Guidelines. UN Doc. A/62/20 of 10 January 2008. Vienna: United Nations. Adopted by: United Nations General Assembly. Resolution on International Cooperation in the Peaceful Uses of Outer Space. UN Doc. A/62/217 of 1 February 2008. New York: United Nations.
[124] Article VI, Draft Treaty on the Prevention of the Placement of Weapons in Outer Space, the Threat or Use of Force Against Space Objects.
[125] Article I (c), Draft Treaty on the Prevention of the Placement of Weapons in Outer Space, the Threat or Use of Force Against Space Objects.
[126] Article I (d), Draft Treaty on the Prevention of the Placement of Weapons in Outer Space, the Threat or Use of Force Against Space Objects.
[127] Considered to be customary international law as well as to be found the Article 2(4) of the Charter of the United Nations – See Nicaragua Case. Case Concerning Military and Paramilitary Activities in and Against Nicaragua. I.C.J. Reports 186, p. 14 at para. 176.
[128] International Civil Aviation Organization.
[129] The location of discussions of the Chinese-Russian treaty proposal at the United Nations Conference on Disarmament.

5.4 Space traffic control: issues and options

William Ailor

Abstract

The first aircraft flew in 1903 and 41 years later, air traffic reached the point where the world recognised the need to manage and control air traffic. Nations worked together to define the right of each nation to control its air space and created the International Civil Aviation Organization to "assure the safe, orderly and economic development of international air transport."[130] As a result, nations developed effective measures to control air traffic entering, leaving, and within their borders.

The first man-made object was placed into Earth orbit in 1957, and 50 years later, problems similar to those that led to a worldwide air traffic control system are arising in space. Space objects are interfering with each other both physically, by posing the possibility of collision, and electronically, by interfering with communications. Some believe that it is time to develop a service to prevent such interference among orbiting objects.

But operations of spacecraft in orbit around Earth are different than aircraft operations, and these differences create challenging issues that must be addressed and compromises that must be made in order to develop and operate an effective space traffic control service. This paper describes the current space environment, what the future may hold, primary stakeholders as space traffic control evolves, issues that must be faced and options for solutions.

5.4.1. What's the problem

When Sputnik was launched in 1957, there were no concerns about there being interference with other objects orbiting Earth – there was only one, the Moon, and it was very far away. As Figure 8 shows, mankind has placed thousand of objects into space since that time. The peaks in the debris curve relate to debris created by the noted events: the explosion of the Ariane V16 in 1986, the explosion of a Pegasus stage in 1996, and the most recent events: the early 2007 Chinese anti-

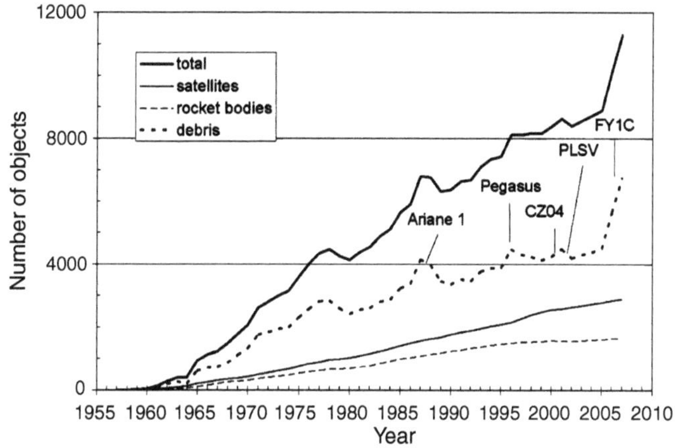

Fig. 8. *Number of objects launched (source: The Aerospace Corporation).*

satellite test that destroyed the Chinese Feng Yung (FY1C) weather satellite and the explosion of a Russian Briz-M rocket booster.

We currently have over 800 operating satellites and over 12,000 other tracked objects, objects larger than 10-cm in diameter, all sharing the same near-Earth space. In addition to these tracked objects, there are hundreds of thousands of smaller, untracked explosion fragments and other objects that also pose a threat to satellite operations. Nations have been working together for years to develop guidelines and regulations to minimise the creation of new debris objects (see Figure 8).

Collisions among these objects are becoming a growing concern for two reasons. First, a collision could terminate a satellite's mission, potentially costing its owner (or his insurance provider) millions of dollars in losses. Second, a collision could create a cloud of debris, some of which could remain a threat to other objects for decades or centuries. It should be noted that orbital speeds, even small particles pose a threat to a satellite's optical sensors and electronics, possibly pitting a critical mirror, or causing damage to a critical component.

To date, we have had three known collisions involving tracked objects. Two of these involved debris objects, and the third severed a boom on an operational satellite, but some functions were later recovered. As Figure 10 shows, collisions are expected to become more common in the future.[131] As an example of what a collision can do, the Chinese anti-satellite test in 2007 created over 2000 new tracked objects and many thousands of untracked fragments.

Operating satellites can interfere in ways besides collisions. For example, satellites broadcasting on similar frequencies can interfere with communications, potentially leading to a disruption of service. The non-profit Satellite Users

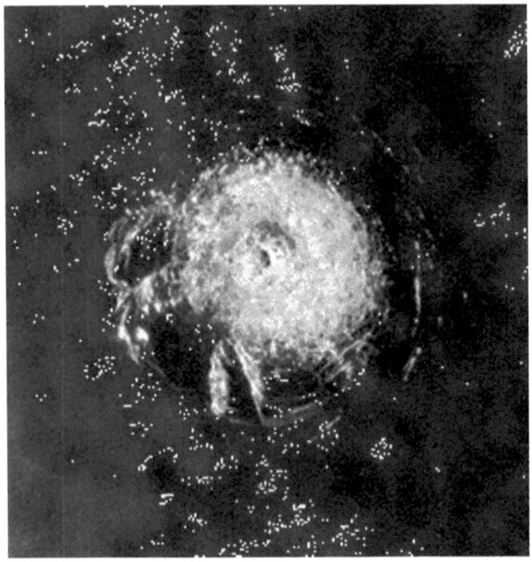

Fig. 9. *Space debris impact (source: NASA).*

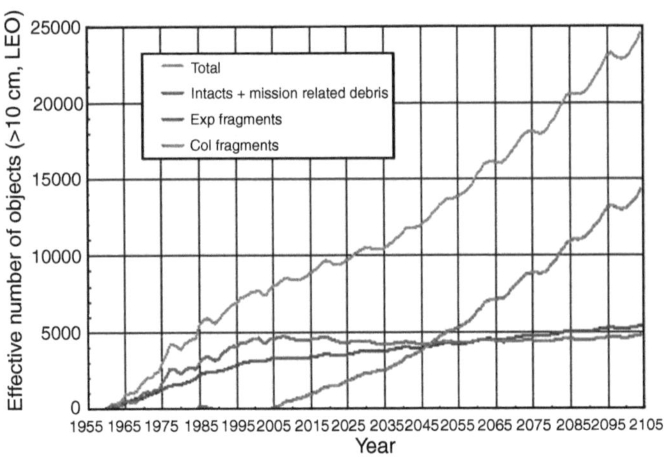

Fig. 10. *Effective number of objects (source: Acta Astronautica, J.C. Liou).*

Interference Reduction Group, SUIRG, was established in 2003 to help resolve interference issues. SUIRG currently has a membership of over 34 organisations, including 20 commercial satellite operators (see Figures 9 and 10).

Given the current and projected near-Earth space environment, it is clear that all satellite operators are increasingly calling for assistance to help them avoid collisions with tracked objects, both debris objects and other operating satellites, and other interference. Manned space vehicles are currently provided such assistance, and the International Space Station and the Space Shuttle have been manoeuvred away from approaches that were projected to be too close for comfort.

While airspace over a nation is controlled by that nation, no such control is possible for objects in orbit, since an object in orbit will eventually pass over all nations within their inclination range. As a result, we need a new form of control that recognises the following basic principles:

1. Collisions in space are bad for every space user and must be avoided.
2. All objects in an orbital regime share virtually the same space.
3. Actions of a single operator have the potential to affect the operations of every other satellite sharing that orbital regime.

Another difference with aircraft control is that non-functional objects share orbital space with operating satellites – as noted earlier, the non-functional objects outnumber the operating satellites by a factor of approximately 15. An even larger number of small, untracked objects increase the risk. The traffic control service must also help operators avoid as many of these risk objects as possible.

5.4.2. Who are the major players?

The major players in space are as follows:

1. Governments who operate lots of satellites, track orbiting objects, maintain catalogs of orbiting objects that enable predictions to be made of where objects will be in the future and also impose operational and disposal regulations on satellite operators.
2. Commercial satellite operators who are operating an increasing percentage of the 800+ operating satellites.
3. Commercial companies that market software tools and tracking data.
4. Non-profit corporations that have prototyped conjunction assessment services.

All of these entities have special needs that provide opportunities for cooperation. For example, while government resident space object catalogues allow reasonable predictions of where non-manoeuvring space objects will be in the future, they do not include information on planned satellite manoeuvres that might affect future positions. This means that catalogues, based on visual or radar observations at specific times, are useful for estimating potential collisions of debris objects and non-manoeuvrable satellites such as the Hubble Space Telescope, but may not be accurate if one or both of the objects under consideration manoeuvres prior to impact.

Unfortunately, many satellites do manoeuvre regularly as part of their missions. For example, satellites in geosynchronous orbits have assigned locations and they are required to operate within specific three-dimensional tolerances of these locations. As a result, satellites periodically execute small corrections to remain within their assigned "boxes". In addition, satellites may occasionally move from one assigned location to another or may move if being inserted into an operational slot or disposed.

Non-profit corporations and commercial companies have already developed and prototyped many of the tools required to initiate an effective service. The experiences of the non-profits should be captured and used as a foundation for future services, and commercial software and tracking data providers should also have a major role to play as the service evolves.

5.4.3. What is required?

Satellite operators require sufficient information to enable them to anticipate close approaches and other events that might affect the operations of their satellites, and this information must be delivered in a timely manner; i.e., with enough warning that they can plan a manoeuvre to minimise the impact on their basic mission and on the long-term sustainability of their operations.

Minimising the impact on their basic mission means that, if required, the operator must be able to plan a manoeuvre that will not affect the satellite's mission. For example, the manoeuvre could simply be a variation of a normal station-keeping manoeuvre rather than a manoeuvre that would move the satellite out of its station-keeping box.

The second requirement, minimising the effect on the long-term sustainability of the satellite's operations, means that the manoeuvre, if required, must minimise the propellant consumed. Propellant is a critical factor in determining a satellite's lifetime and operators must manage this resource carefully.

The above requirements also translate to a need for accurate data. Studies show that basing collision avoidance moves on inaccurate data leads to more manoeuvres. If the desire is to reduce the 10-year risk of collision for a satellite in geosynchronous orbit by a factor of 10, over 10 moves per satellite per year are required if General Perturbations data (the lowest resolution data generally available to the public) is used for predictions.[132] If the highest resolution data are used, the number of manoeuvres drops to less than one per year.

Since space is shared by multiple operators, both public and private, an effective hazard avoidance service must assure that all satellites are covered. This means that data on space junk, government satellites and satellites operated by private companies must be included.

Operators also desire the service to have other characteristics. For example, if an operator is notified of a potential problem and develops a manoeuvre plan to minimise the hazard, the operator wants to be sure that the manoeuvre doesn't increase the probability of another close encounter. Also, before a manoeuvre is actually executed, the operator may wish confirmation that tracking data on the approaching object has been updated and that the hazard still exists.

If the approaching object is an operating satellite, the avoidance manoeuvres must be coordinated to avoid moves by both operators which could actually worsen the problem. Also, for repeat conjunctions, conjunctions which occur on a periodic basis, operators may wish to work together to ensure adequate ongoing separation. The service might include assistance in opening communications among affected operators.

Summarising, satellite operators require as follows:

1. Timely warnings of potential threats to normal operations.
2. Predictions based on as accurate data as possible.
3. Predictions that include ALL orbiting objects – space debris, government satellites and private satellites.
4. Assistance in verifying that planned manoeuvres will actually reduce the probability of interference.
5. Updated tracking data to verify that interference may occur.
6. Assistance in contacting operators of satellites with whom interference is predicted.

Finally, operators desire a reliable service that is available at all times and is "guaranteed"; i.e., that an operator knows what information and assistance will be delivered when. Spacecraft orbits are independent of the world situation and political differences among nations. A space traffic control service must be available to (and must ideally include) all satellite operators.

5.4.4. What are the issues?

Virtually all satellite operators recognise the need for a service to assist with minimising the possibility of interference, but all recognise that there are issues that will affect how such a service is structured. Some of these issues are as follows:

1. Governments control the best tracking data on satellites, but the government agencies that have this data are not chartered to provide support to non-government operators. Will a government be willing to provide services to commercial operators and operators of satellites by other governments? Will that government be willing to guarantee a level of service? Will that level of service be satisfactory to non-government operators?
2. Commercial satellite operators have the best data on their own satellites, but who are commercial operators willing to trust with that data?
3. Operators want a guaranteed level of service. Is there an organisational structure where this can be assured?
4. Do we have the tools required to provide a service, or is research and development required?
5. To date, we've not had a significant collision in space, one that has affected critical operations of a valued asset. How will this fact affect the evolution of a space traffic control service?
6. Who will pay for a space traffic control system?
7. Is there an organisational structure that offers characteristics that encourage governments and private satellite operators to participate?

5.4.5. Options for moving forward

5.4.5.1. Government provider

As noted, governments maintain the most complete catalogues of orbiting objects. Governments have also developed and maintained the infrastructure of optical and radar tracking resources required to monitor and update their catalogues. These resources have been built with government or taxpayer funds in individual nations, and their use to provide support to satellite operators outside of those nations' responsibility may be difficult to justify.

In addition, just as is the case with private operators, governments may not wish to make catalogues that contain information on sensitive satellites or other information generally available or to provide this information to one or two governments offering to provide a space traffic control service.

So how might governments participate in a space traffic control service? One option is for a single government (e.g., the U.S. or Russia) or group of governments (e.g., ESA) to offer to take responsibility for providing such a service. In doing so, the government would need to resolve issues raised earlier about how government resources would be paid for, what agency would provide the service, etc. Perhaps the more difficult tasks would be to convince other governments and private operators that the service could be trusted to properly protect sensitive data, that the service would represent all operators worldwide, that the level of service would be what operators require, and that the service would be guaranteed no matter the world situation.

5.4.5.2. Private for-profit company

A private, for-profit company could announce its intention to take this responsibility, but there are some major hurtles. First, since there has been no major collision to date, the market for services may be very limited, which may limit the amount that could be charged to users. Second, the organisation would require tracking and catalog data, and governments may be reluctant to provide their best data to an entity that will make a profit from that data. Third, liability should a collision occur could be a significant risk for a commercial company. Finally, for-profit companies may not offer the stability required for the long term. These companies can be bought and sold, so that ownership and protection of proprietary and sensitive data might be viewed as a concern.

5.4.5.3. Non-profit company

Another option is to create an international entity, perhaps an international non-profit organisation that would provide services by melding data from a variety of sources, both government catalogues and private operators.[133,134] According to Wikipedia: "a non-profit organisation (abbreviated "NPO", also "not-for-profit") is a legally constituted organisation whose objective is to support or engage in activities of public or private interest without any commercial or monetary profit."[135] Continuing, "The extent to which [a NPO] can generate income may be constrained in amount, methods or both, and the use of those profits may be restricted not only in purpose but in proportions regarding self-maintenance and achievement of purpose. NPOs therefore are typically funded by donations from the private, public sector, or both, as well as from programme service fees".

An NPO providing space traffic control services could be governed by a board of directors made up of representatives of both governments and private satellite operators – the government representatives assuring that any data provided by their governments is handled appropriately and the operator representatives assuring that they get the services they need.

Since governments are likely to continue to perform their own hazard avoidance analyses for sensitive objects, a key requirement for governments will be to receive the best data available on private satellite operators and planned manoeuvres. With agreement of the private operators, the NPO could provide this data in trade for the best catalog data.

Private operators could contract with the NPO for desired services, with the fee for services set by the board of directors. The NPO could also contract with tracking services for special tasking of sensors to resolve close approaches.

A critical feature of an NPO is its ability to work as a partner with governments. This feature may make it much easier for governments to share catalog data with the non-profit. And the board of directors may require that all "profits" made by the organisation go to improving its services and products.

Since preventing interference is of interest to all space operators, financial support for the organisation would come from both governments and private operators. The organisation would have duplicate physical and computer facilities operating in separate geographic locations to ensure ongoing operations in the event of a fire or other emergency at one site.

Finally, the organisation would not be operated by a single government, but by a board representing multiple nations and private operators. This feature may help the organisation evolve quickly.

5.4.6. Moving forward

A fortunate feature of space traffic control is that-technically, the job can be done now. As noted, manned systems in low, Earth orbit are already monitored for potentially threatening approaches by other objects. The Aerospace Corporation and others have provided similar services to operators in geosynchronous orbits, and software for predicting and visualising close approaches is commercially available. A number of private geosynchronous satellite operators have created a voluntary consortium where operators, with the assistance of a commercial software company, share data and look for interference.[136,137] So the basic tools and capabilities are documented and available.

Space traffic control has not moved forward at a higher, more formal level because (1) there has been no major loss of a space system due to collision and (2) an organisational structure that deals with the significant political and service issues noted earlier has not been put forward as a serious option.

The fact that a good space traffic control service requires nations and private operators to work together poses serious challenges. A properly structured non-governmental organisation appears to have characteristics that protect the multiple interests involved. Hopefully, significant progress can be made before a collision that might have been preventable demonstrates the need for effective space traffic control.

[130] Pelsser, Albert. "ICAO Emblem and its History." ICAO. 5 Nov. 2009. http://www.icao.int/cgi/goto_m.pl?icao/en/icao_emblem_history.pdf.
[131] Liou, Jer-Chyi."A Statistical Analysis of the Future Debris Environment." Acta Astronautica 62 (2008): 264–271.
[132] Ailor, William.H. and Glenn E. Peterson. "Collision Avoidance as a Debris Mitigation Measure." IAC-04-IAA.5.12.3.01. 55th International Astronautical Congress. Vancouver, B.C., Oct. 2004.
[133] Ailor, WilliamH. "Moving Forward on Space Traffic Control." Presentation. 3rd International Association for the Advancement of Space Safety Conference. Rome, Italy. Oct. 2008.
[134] Ailor, WilliamH. "Space Traffic Control: The Nonprofit Option." IAC-08-E3.2.2. 59th International Astronautical Congress. Glasgow, Scotland, Sept. 2008.
[135] "Non-Profit Corporation." 2009. Wikipedia. 5 Nov. 2009. http://en.wikipedia.org/wiki/Non-profit_organisation.
[136] Dal Bello, R. and J. Chan."Data Sharing to Improve Close Approach Monitoring and Safety of Flight." Presentation. 3rd International Association for the Advancement of Space Safety Conference. Rome, Italy. Oct. 2008.
[137] Kelso, T.S., Valado, D.A. Chan, J., Buckwalter, B. "Improved Conjunction Analysis via Collaborative Space Situational Awareness." Presentation. 3rd International Association for the Advancement of Space Safety Conference. Rome, Italy. Oct. 2008.

5.5 The concept of space traffic management as a basis for achieving the fair and equitable use of outer space

Kai-Uwe Schrogl

Fairness and equitability in the use of outer space can be reached in various ways. They could be approached through single steps and actions, or they could be approached through applying a comprehensive new concept for regulating space activities. Such a concept is Space Traffic Management (STM). STM received its first full conceptualisation in a study published in 2006 by the International Academy of Astronautics (IAA).[138] Following this, numerous academic studies as well as practitioners' activities have been undertaken to further refine and give substance to this concept. In parallel, major policy initiatives (the EU Draft Code of Conduct, the proposal by the former UNCOPUOS Chairman Gerard Brachet on the sustainable use of outer space and the proposal by the current UNCOPUOS Chairman Ciro Arévalo Yepes on a UN space policy) have addressed the content of STM. This article will set out the policy and legal framework for STM[139] and provide links to the current policy initiatives. It further explains how STM can be used as the framework and enabler for introducing elements of fairness and equitability in the use of outer space.

5.5.1. The concept of "traffic" in outer space

Compared with road traffic or air traffic, it seems bold to use the term "space traffic". At first glance there are no congested roads where a variety of traffic participants fight for their rights and there are no congested runways where management of take-offs and landings is done by the minute. In fact, since the beginning of the space age only 30,000 man-made objects larger than 10 cm have been observed and registered. Today there are 12,000 objects in earth orbits, 1100 of which are in the geostationary Earth orbit (GEO). Given the relative physical dimensions, this does not sound very dramatic. But viewing "space traffic" from a different angle changes this initial evaluation. The first aspect to be considered in more detail is the distribution of space objects. In fact, there are areas with a particularly high density of activities: the low orbits of up to 400 km, the polar orbits at around 800–1000 km and the GEO at 35,800 km above sea level. With some notable exceptions (such as medium Earth orbits where navigation satellites can be found), these orbits host the largest number of the main applications satellites, such as those for Earth observation and telecommunication. This uneven distribution has already led to forms of congestion – a fact that has been known for years with regard to the GEO.

Currently, at least 800 active satellites are in orbit (around half of them in the GEO), moving with typical velocities of up to 7500 m per second. Only a few of them have manoeuvring capabilities. They are surrounded by a growing amount of space debris. Currently only objects larger than 10 cm can be tracked in all orbits, with a total number greater than 10,000. However, the millions of smaller objects (larger than 1 cm) can still do harm to satellites (or humans in outer space). This debris population is constantly increasing, in particular through explosions of upper stages, which happen on average at a rate of five per year. The Chinese anti-satellite test of January 2007 additionally created a population of more than 2000 trackable pieces of debris in a highly valuable orbit plain where debris remains for decades or centuries. Three collisions between active satellites and space debris have already been recorded, and the need to fly debris avoidance manoeuvres by the Space Shuttle and the International Space Station has become routine. This second perspective on space traffic draws a more urgent picture of the situation around the Earth. This is why research on how to cope with the problem of maintaining space for safe use has been on the rise for the past 2 years.

5.5.2. Elements of STM: findings from the 2006 study by the International Academy of Astronautics

In the early 1980s, the term "traffic" started to be used for space activities and the resulting need for regulation.[140] In a more comprehensive way, the American Institute of Aeronautics and Astronautics (AIAA) took up the issue at two workshops they organised at the turn of the century.[141] Emanating from these workshops, the International Academy of Astronautics (IAA) established a working group on the issue of Space Traffic Management (STM) in order to prepare an in-depth multi-disciplinary study (a "Cosmic Study" in the IAA's nomenclature). This study was published in 2006. It is the first comprehensive work in this field.[142]

Fig. 11. *IAA cosmic study on Space Traffic Management (source: IAA).*

The study[143] defines STM as "the set of technical and regulatory provisions for promoting safe access into outer space, operations in outer space and return from outer space to Earth free from physical and radio-frequency interference". The study acknowledges that the need for STM is not yet so pressing that immediate action has to be taken. But it clearly identifies perspectives that suggest it would be prudent to start now to conceptualise a future regime. While the study refers to a slow and steady decline of launches since 1980, it nonetheless stresses the likely implications of the growing number of countries with their own launching capacities and launch facilities. The number of catalogued objects is also rising due to space debris, although the number of active satellites remains at 6–7% of the total number of catalogued objects. The precision of current space surveillance systems needs to be enhanced and data sharing has to be developed further. Also, information on "space weather" is still limited and must improve.

The study also finds that the prospects for re-usable space transportation systems are still open, that human space flight will roughly remain at 10–15% of all launches (a proportion which has been constant over the past 20 years) and that following the successful flight of Spaceship One there might be – if safety is guaranteed – a growing number of sub-orbital human flights. The picture is supplemented with a review of novel technologies (such as tethers, stratospheric platforms and space elevators) that might be introduced in the future and will also have to be taken into account. This enumeration of perspectives shows the variety of trends and developments that will make space activities more diverse with respect to both technologies and actors. This poses additional challenges, including in the regulatory field.

An STM regime will comprise four areas: the securing of information needs, a notification system, concrete traffic rules and mechanisms for implementation and control. The first area is the basis for any kind of traffic management in outer space. In order to manage traffic, a sound information basis regarding Space Situational Awareness (SSA) has to be established. Today, only the U.S. Strategic Command possesses such a capability and shares some of its information with external users. A global STM system has to be open and accessible to all actors. The challenge will be to exactly define the necessary data and to establish rules for data provision and data management as well as rules for an information system on space weather. Only on the basis of shared knowledge of what is going on in the Earth orbits, can traffic rules become meaningful.

The second area is a notification system. The current system of registration based on the Registration Convention of 1975 is insufficient by far. A pre-launch notification system together with a notification system on in-orbit manoeuvres has to be established. To this end, parameters for the notification of launches and operations of space objects need to be worked out. This must be complemented by rules for the notification of orbital manoeuvres and for re-entries. In addition, provisions are necessary for the notification of the end-of-lifetime of space objects.

The third area comprises the concrete traffic rules, which – by analogy to road traffic – come first into mind when traffic management in outer space is mentioned. Here there are some direct analogies with terrestrial traffic management, but also completely different rules. This area begins with safety provisions for launches, then turns to space operations with right-of-way rules (comparable to "sail before motor" in maritime traffic), then encompasses prioritisation with regard to manoeuvres, specific rules for the protection of human spaceflight, zoning (e.g., keep-out zones, providing special safety to military space assets[144]), specific rules for the GEO, specific rules for satellite constellations, debris mitigation rules, safety rules for re-entry (e.g., descent corridors) and environmental provisions (e.g., the prevention of pollution of the atmosphere and the troposphere).

The fourth area will have to deal with mechanisms for implementation and control. The "modern" way of law or rule making by international organisations such as the ITU or the International Civil Aviation Organization (ICAO) provides an appropriate precedent that could be followed. Basic provisions could be laid down in an international treaty (either drafted by an ad hoc assembly of States or in the framework of an existing organisation like ICAO or an existing forum like UNCOPUOS), and subsequent rules of the road and standards could be developed in a routine way in the format of soft law. Since space law making in UNCOPUOS is very traditional, such an innovation would be a real culture change.[145] Another innovation would be the introduction of enforcement and arbitration mechanisms ultimately leading to a kind of policing in outer space and sanctions such as the withdrawal of access to information or of the use of frequencies. This might sound utopian to those who adhere to traditional space law, but it is part of the systemic approach taken by STM, which makes it such a revolutionary concept.

Finally, STM touches and sheds new light upon an issue which has been on the agenda for decades. While the idea for a World Space Organisation (WSO) has been around for more than 30 years, so far no convincing answer has been provided to the question of what role such an organisation should play. STM requires strong operative oversight. This could be exerted by a WSO but the authors of the IAA study made it clear that it might be preferable to broaden the mandate of the already existing and efficiently operating ICAO rather than establishing a new big bureaucracy. The philosophy behind this proposal is that space traffic might ultimately (but only in some decades) evolve into air traffic in another dimension where States and private actors with their "spacelines" operate side by side under one regulatory umbrella. But even without this optimistic vision, STM is a timely idea to shape the debate on how to overcome the regulatory deadlock that exists today.

5.5.3. Research on STM following the IAA study

The IAA study was the first comprehensive approach to shape a space traffic management system. Its results have been presented at conferences and published in articles and have been brought to the attention of UNCOPUOS. Dedicated sessions on STM are starting to be held at symposia and at institutions such as the International Space University (ISU) which conducted a student project on this issue in 2007. In the U.S., the Center for Defense Information (CDI) – a Washington based think tank – is also concerned with STM in its research on space security.[146]Another notable initiative in this field was the establishment some years ago of the International Association for the Advancement of Space Safety (IAASS) which in May 2007 published a thorough report in which STM was reflected as a cornerstone for space safety.[147] The European Space Policy Institute (ESPI) has been involved in most of these initiatives of ISU, CDI and IAASS, and will continue to play a leading role in research related to the conceptualisation of STM as well as in research on specific related topics.[148] IAA itself will continue its research in the field. Its Commission on Policy, Economics and Regulation is currently identifying topics for further related study projects. Increased visibility was also generated by private initiatives that aim at establishing STM as a commercial service.

All these follow-ups and initiatives show that the concept of STM has its place in the framework of commemorating the anniversaries of the first space flight and the entering into force of the Outer Space Treaty. Its mission, however, is to show that we have already entered a new era of using outer space. It is an era that is characterised by new technologies and, even more important, by a growing number and type of actors. Preparing for the regulatory "big bang", leading to an effective framework for the safe and equitable use of outer space, may take more than a decade. Therefore, it is encouraging to note that the debate – to a great extent initiated by IAA – is seriously underway.

5.5.4. STM as a new conceptual approach to regulating space activities

While space technologies have changed rapidly, space law has not moved beyond the basis that was set by the Outer Space Treaty 40 years ago.[149] Current space law still shows strong traces of its origins in the East vs. West era. It remains characterised by a primary focus on States as actors in outer space. This has led to a situation where the growing need for effective mechanisms to regulate the activities of non-governmental private actors is not being met. Examples of areas where this has become particularly apparent include reviews of the concept of the "launching State" and registration practice with regard to private actors.[150] While problems have been highlighted through these agenda items and proposals for remedies have been developed, the deliberations have shown, yet again, that UNCOPUOS is a forum that is characterised by an extremely slow decision-making process and a barely imaginable reluctance of its Member States to accept any changes in the current regulatory framework.

The extreme timidity of the Member States of UNCOPUOS has resulted over the past 15 years in the absence of any new binding provisions to international space law. Moreover, there have been no authoritative interpretations of existing provisions, which urgently need to be re-evaluated in the light of new developments. The consequence has been that, on the one hand, other international organisations such as the International Telecommunication Union (ITU) have started to regulate areas of space activities while, on the other hand, soft law (regulations, standards, etc.) instead of binding international law has been developed in technical forums such as the Inter-Agency Space Debris Coordination Committee (IADC), the Committee on Earth Observation Satellites

Fig. 12. *United Nations treaties and principles on outer Space (source: UN).*

(CEOS) and through initiatives like the Hague Code of Conduct Against Ballistic Missile Proliferation (HCOC). Consequently, UNCOPUOS is slowly losing control over the regulation of outer space activities.

In this situation, STM challenges the current condition of space law as well as the way it is being developed. By contrast to the current situation which could be described as "piecemeal engineering", STM would provide a regulatory "big bang". STM would not tackle single issues, but would approach the regulation of space activities as a comprehensive concept. This concept is based on functionality, aiming at the provision of a complete set of rules of the road for the current situation and future developments. Space activities must be viewed as a holistic traffic system rather than the disconnected activities of States. This requires not only new levels of interaction and forms of regulation (binding treaty provisions/ technical standards, international/national provisions) but also new ways of organising supervision and implementation.

It is clear that specific provisions of current space law can and will find their way into such a regime. This will certainly be the case for principles such as freedom of use, non-appropriation and peaceful uses. A comprehensive approach would also make it possible to integrate existing regimes for specific areas, such as the ITU regulations on using the orbit/frequency spectrum of the GEO, the emerging space debris mitigation regime developed in IADC or even the rocket pre-launch notification regime of the HCOC (which is one more area where UNCOPUOS with its post launch registration regime has been outpaced). But STM would be more than merely the sum of these single parts. STM would be able to develop all provisions in a coherent way based on the overarching principle of guaranteeing safe operations in the space traffic system.

5.5.5. STM and its relation to current policy activities

Establishing such a comprehensive STM regime is, however, a task for the more distant future. It would require an effort comparable to the decade-long global negotiation process of the new Law of the Sea up to the 1980s. Initial approaches for restricted issues related to STM are already being taken. Four of them are particularly remarkable and will be briefly presented in this section.

In December 2008, the EU Council Presidency presented a Draft Code of Conduct (CoC) for space activities.[151] This initiative emerged from the frustrating deliberations in the Conference on Disarmament but it is not intended to be discussed there. The Draft CoC was the first governmental position statement on the issue of regulating space activities in a way that can be regarded as comprising elements of an STM regime. It comprises notification and registration and also provisions for space operations. While it is a relatively short document, it has opened the door for serious discussions on managing activities in outer space beyond those foreseen in the current legal regime, leading to STM as outlined above as a regulatory "big bang". The EU Council Presidencies, which have been involved in the preparation of the Draft CoC since 2007, had been briefed on the 2006 IAA study on STM in order to find a basis for their initiative there.

In 2007, the Chairman of UNCOPUOS, Gerard Brachet, referring to the results from the IAA study, suggested putting STM on the Committee's agenda.[152] After further refinement of his proposal, in 2009 the Committee adopted a new agenda item "Long-term sustainability of outer space activities", which will be dealt with under a multi-year work plan in the Scientific and Technical Subcommittee between 2010 and 2013. The goal is the preparation of a report on the long-term sustainability of outer space activities and the examination of measures that could enhance it, together with the preparation of a set of best practice guidelines. It is even foreseen that the Legal Subcommittee might become involved and that such best practice guidelines might develop into a UN General Assembly Resolution. This effort will certainly lead to the identification of elements for a STM regime, but can only be regarded as a first step to ascertain the interests of the States in this field and to prepare common ground in the practical technical area.

In parallel to this proposal, which will start to formally seek concrete results in 2010, the current Chairman of UNCOPUOS, Ciro Arévalo Yepes, has proposed establishing a "UN Space Policy"[153] which also contains reference to the UN's role "to regulate the orbital environment for the fair and responsible use of space". This initiative supports the work plan on sustainable use in that it seeks to establish a policy framework. Other than UNCOPUOS, as early as 2005, the then president of the ICAO Council proposed that the organisation should think about the issue and start to play a role in regulating space activities.[154]

These four initiatives show that the management of space operations became an inter-governmental issue shortly after the IAA study on STM was published in 2006. The initiatives described above will now go their rather slow and formal ways in international negotiating processes or institutional formats such as the multi-year work plan in UNCOPUOS. However, they only cover single elements and do not yet provide a comprehensive approach for a new regulatory concept based on STM. This may emerge only after first experiences with the outcomes of the current initiatives in one or two decades.

5.5.6. STM as an approach to conceptualise equitable and fair use of outer space

Fairness and equitability in the use of outer space[155] are already enshrined in specific provisions of the Outer Space Treaty (e.g., Art. I) and in single issue regimes such as the use of the GEO (e.g. the a priori planning of 1988). But a broader and more coherent conceptualisation has not been achieved and cannot be achieved on the basis of the current heterogeneous settings in space law. Single elements of fairness and equitability are contained in specific provisions of the initiatives presented in Section 5.5.5. A thorough approach, however, can only be achieved through an equally comprehensive concept for regulating space activities. And this is the case with STM. It not only provides the framework for establishing concrete rules, it also provides the enabling mechanism for applying fairness and equitability throughout.

STM can respond to questions like: Who are the actors demanding equitable treatment and fairness (are they only States or are they also non-governmental entities)? How is access to space organised? How are the rules in orbit shaped? How are the benefits of space applications and exploitation distributed? How is cooperation and coordination organised? How is information provided and shared? How is ownership of space resources organised? These are only a few of the basic questions that can be approached in a new and fresh way when considering a comprehensive STM regime in and for the future. Unfettered by existing rules but basing its considerations on past and current experiences, the drafting and negotiating process for STM is open and available to achieve a new level of fairness and equitability in the use of space.

[138] Contant-Jorgensen, Corinne (Secretary of the Study Group), Lala, Petr and Kai-Uwe Schrogl (Coordinators of the Study Group), eds. Cosmic Study on Space Traffic Management. Paris: International Academy of Astronautics, 2006. The Study Group consisted of 16 contributors from numerous countries covering engineering, policy and legal aspects. Download at: http://iaaweb.org/iaa/Studies/spacetraffic.pdf.

[139] The article by Bill Ailor in this book focuses more on the technical aspects of STM.

[140] Perek, Lubos. "Traffic Rules for Outer Space." in: Proceedings of the Twenty-Fifth International Colloquium on the Law of Outer Space, 27 Sept.–2 Oct. 1982, Paris, France. Ed. International Institute of Space Law (IISL). 82-IISL-09. Reston, VA: American Institute of Aeronautics and Astronautics, 1983.

[141] American Institute of Aeronautics and Astronautics (AIAA), ed. Proceedings of the 5th International Space Cooperation Workshop "International Cooperation: Solving Global Problems." April 1999, Bermuda. Reston, VA: American Institute of Aeronautics and Astronautics, 1999. 35–39; American Institute of Aeronautics and Astronautics (AIAA), ed. Proceedings of the 6th International Space Cooperation Workshop "International Cooperation: Addressing Challenges for the New Millennium." March 2001, Spain. Reston, VA: American Institute of Aeronautics and Astronautics, 2001. 7–14.

[142] Contant-Jorgensen, Corinne (Secretary of the Study Group), Lala, Petr and Kai-Uwe Schrogl (Coordinators of the Study Group), eds. Cosmic Study on Space Traffic Management. Paris: International Academy of Astronautics, 2006. The Study Group consisted of 16 contributors from numerous countries covering engineering, policy and legal aspects. Download at: http://iaaweb.org/iaa/Studies/spacetraffic.pdf.[143] For the following see Schrogl, Kai-Uwe. "Space Traffic Management: The new comprehensive approach for regulating the use of outer space. Results from the 2006 IAA cosmic study."Acta Astronautica 62 (2008): 272–276.

[144] Such keep-out zones could also be a topic for the blocked negotiations in the Geneva Conference on Disarmament's Committee on the Prevention of an Arms Race in Outer Space (PAROS). Since the threat to military space assets is one of the drivers for a possible weaponisation of outer space, STM could through such specific means also contribute to arms control.

[145] Early ideas on such an approach by Jasentuliyana, Nandasiri. "Strengthening International Space Law." in: Proceedings of the Third ECSL Colloquium on "International Organisations and Space Law", 6–7 May 1999, Perugia, Italy. Ed. European Centre for Space Law. esa SP-442. Paris: European Centre for Space Law 1999. 87–96.

[146] See the activities of the CDI's Space Security Program, led by Theresa Hitchens, at http://www.cdi.org/program/index.cfm?programid=68.

[147] International Association for the Advancement of Space Safety (IAASS). "An ICAO For Space?", 2007. Download at: http://www.iaass.org/pdf/ICAO%20for%20Space%20-%20White-20Paper%20-%20draft%2029%20May%202007.pdf.

[148] See Rathgeber, Wolfgang. "Europe's Way to Space Situational Awareness (SSA)." 10 Jan. 2008. ESPI Report 10. European Space Policy Institute 11 Aug. 2009. http://www.espi.or.at/images/stories/dokumente/studies/ssa.pdf., which deals with data policy issues related to SSA.

[149] See the recently published first comprehensive commentary on the Outer Space Treaty: Hobe, Stephan, Schmidt-Tedd, Bernhard and Kai-Uwe Schrogl (eds.), Gerardine Goh (ass. ed.). Cologne Commentary on Space Law, vol. 1: Outer Space Treaty. Cologne: Heymanns, 2009.

[150] Both topics have recently been dealt with in working groups under multi-year work plans of the UNCOPUOS. See: Schrogl, Kai-Uwe and Charles Davies. "A New Look at the Concept of the "Launching State". The Results of the UNCOPUOS Legal Subcommittee Working Group 2000-2002." German Journal of Air and Space Law ZLW 51.3 (2002): 359–381. and Schrogl, Kai-Uwe and Niklas Hedman. "The U.N. General Assembly Resolution 62/101 of 17 December 2007 on "Recommendations on Enhancing the Practice of States and International Organizations in Registering Space Objects." " Journal of Space Law 34.1 (2008): 141–161. The author has been the chairman of both these working groups.

[151] See Rathgeber, Wolfgang and Nina-Louisa Remuß. "Space Security – A Formative Role and Principled Identity for Europe." 16 January 2009. ESPI Report 16. European Space Policy Institute. 11 Aug. 2009. http://www.espi.or.at/images/stories/dokumente/studies/espi%20report%2016.pdf: 58–64.

[152] Brachet, Gérard. Future role and activities of the COPUOS. Working paper submitted by the Chairman. UN Doc.A/AC.105/L.268 of 10 May 2007. Vienna: United Nations. para. 28.

[153] Arévalo Yepes, Ciro. "Towards a UN Space Policy." 23 June 2009. ESPI Perspectives 23. European Space Policy Institute. 11 Aug. 2009.http://www.espi.or.at/images/stories/dokumente/Perspectives/espi%20perspectives%2023.pdf.

[154] See van Fenema, Peter. "Suborbital Flights and ICAO." Air and Space Law 30.6 (2005): 396–411.

[155] See the contribution to this book by Wolfgang Rathgeber on the conceptualisation of fairness.

6 Achieving global engagement – summarising introduction

Gerard Brachet

Session 3, devoted to the topic "Achieving global engagement", began with a presentation by Xavier Pasco (FRS, France) on overcoming international obstacles. The gist of his presentation was to stress that States' space activities are primarily driven by political considerations, not market forces, and therefore tend to amplify power differences. At the international level, this translates into some States being members of the "space-faring nations club", which defines the space security political landscape, while other nations who are outside of the club benefit from the development of space-based services but do not participate in shaping their political and legal framework. Nations who are members of the space club share some common concerns, particularly the security and safety of space operations (space debris, space weather), while nations outside the club stress that rulemaking has to take account of the overall interests of nations and international organisations using space assets. As a consequence, only win–win scenarios are acceptable, where both members and non-members of the space club benefit equally from "rules of the road" aimed at a fair and responsible use of space. A good deal of institutional creativity will be needed to achieve this.

The second speaker, Driss El Hadani (CRTS, Morocco), addressed the question "How to engage present and future actors" He stressed that the increasing diversity of actors in space (47 States have spacecraft in orbit), the emergence of a new category of space-faring nations (India, China, Brazil, etc.), and the development of commercial operators and commercial uses of space systems, is leading to new threats to the orbital environment. The interest of the developing world in space applications for their economic and social development, particularly in civil security, translates into an increasing awareness that space policymaking is important for them. Therefore, fair and responsible use of space goes much beyond the issue of space assets and their security – it now includes all applications of space systems to Earth-bound problems. A strong political commitment by the actors is essential to make any progress (e.g., the EU–Africa partnership, the GEOSS process). Developing nations will need to build a well-established space policy/strategy going much beyond the single project approach, and will need strong regional organisations, as well as encouraging the involvement of the private sector. Easier access to orbital information, to scientific data and to environmental data is a prerequisite for developing nations to play an active role. As for international cooperation, clear vision translating into clearly stated objectives, combined with strong cooperation models, preferably based on existing structures, will be required. Any new institutional mechanism set in place to ensure a fair and responsible use of space will need to be based on a shared "ethic" of space activities, not on market-driven rules.

The third speaker, Professor Kazuto Suzuki (University of Hokkaido, Japan), addressed the role of international organisations for a fair and responsible use of space. He started with an interesting series of theoretical concerns based on some traditional assumptions about international relations and organisations which are not entirely true in outer space activities: the actual behaviour of States is sometimes not consistent with their international obligations. In addition, there are limitations to what an international organisation can achieve if its Members States do not implement its rulemaking. After reviewing the authority, the "power resource", that can be vested in an international organisation, including its agenda-setting and norm-making authority, Prof. Suzuki discussed the limits of this power: How to convince Member States to abide by its rules when it is against their perceived best interest? How to encourage States that are not members of an international organisation to follow its rules and norms? How to give an international organisation binding power, including sanctions? According to Prof. Suzuki, international organisations can overcome their limitations if one Member State, or a group of Member States, exercises strong leadership and if it builds on similar international norms developed elsewhere. For outer space activities that are of interest to all States, developed or emerging, as well as to civil, military and commercial operators, a strong moral role can be played by an international organisation. With respect to establishing an international norm for a fair and responsible use of space, the attitude of major space-faring nations will be key for its credibility, but it will also need to be of mutual benefit to all actors, with an appropriate balance between obligations and sanctions.

The fourth speaker, Ms. Agnieszka Lukaszczyk, Secretary General of the Space Generation Advisory Council (SGAC), addressed the issue of how to involve the world's youth. The creation of the SGAC in 1999 during the Unispace III Conference in Vienna is a good example of a wide-ranging action to involve the younger generation worldwide. Its vision is "Advancing human development through peaceful use of Outer Space". SGAC has members from 87 countries who are actively involved in interactive discussions on issues affecting the future of space activities such as space debris, space traffic management, governance and conflict avoidance. She concluded by presenting some recommendations developed by the

SGAC including one on establishing a set of "Rules of the Road" to safeguard outer space and one on "Space Governance" (specifically lunar governance and property rights).

The discussions that followed the four presentations first addressed the role of private and commercial actors. It was recognised that the private sector is playing an increasingly important role acting within the rules and regulations agreed upon by States at the international level. In outer space, States are ultimately responsible for the acts of private companies under their jurisdiction. On this particular issue, however, Theresa Hitchens pointed out that the term "responsibility" (of States) is not understood everywhere as having the same meaning. It was pointed out that the increasing role of the commercial sector in outer space could also be recognised through mechanisms of representation similar to those introduced many years ago by the ITU in the area of telecommunications. The issue of increasing the role of COPUOS and of modernising its mode of operation was addressed. Driss El Hadani made several suggestions as to how to enhance participation of Member States, increase resources available to the Secretariat (UNOOSA) and improve the efficiency of the annual inter-agency meeting. Ambassador Ciro Arevalo emphasised the recent rejuvenation of COPUOS and his intention to push for a more proactive COPUOS during his chairmanship of the committee. The notion of the "club" of space-faring nations was also discussed. As space applications develop and touch upon so many aspects of modern society, all nations are interested in the future of space-based services. In addition, as Ambassador Arevalo stressed, many developing countries are keen to master not only the user terminals and software of space applications, but also to learn to master the hardware of space systems, hence the increasing number of less developed nations who buy and operate their own satellite systems.

6.1 How to engage current and future space actors

Driss El Hadani

6.1.1. Introduction

Since the launch of the first artificial Earth satellite, space activities have been characterised by a situation of strategic competition between space powers based on geopolitical considerations. Despite this competition, during the last 50 years international cooperation has played an important role in the development of space uses and utilisations. Nevertheless, over the past decade the emergence of new actors, geopolitical developments and technological advances have shaped a new landscape. The changing international environment (greater risks and more competition) is impacting the way nations use space. Several factors have made space more important for national security including: the international influence that space activities can bring, the technological capabilities that space programmes allow nations to acquire and the informational advantages that space assets and services bring to security operations. Space actors are the international community (current and future) and they are facing new challenges to achieve a global commitment for a responsible and fair use of space.

6.1.2. The context

Space is the driving force behind technological progress. From the beginning of the space age, it was considered to be a strategic arena. The end of the Cold-War dramatically cleared the way for States to engage in even greater international cooperation in space activities. But the significant political developments that have occurred at the beginning of the 21st century have led several States to increasingly view the "peaceful" use of space as being strategic especially to their security. This new context could reduce the number of opportunities for cooperation even if States continue to consistently emphasise the importance of international cooperation. This situation presents us with a paradox. While the benefits of cooperation in the economic development of outer space continue to increase, so does the risk of "weaponisation". The evolution of this cooperative competition will impact on space cooperation and security depending on whether or not States pursue independent or collective measures to achieve the global goals set out in their space policies.

6.1.2.1. Diversity increases among stakeholders in space

In 1990, we witnessed a dramatic increase in the rate at which new States saw their demands satisfied and gained access to space. According to the "Space Security 2008" report, "...by 2007, 10 actors had demonstrated independent orbital launch capacity and 47 States had launched civil satellites, either independently or in cooperation with others". The increasingly strong global interest in space has become an important tool for development: Algeria, Brazil, Chile, Egypt, Malaysia, Morocco, Nigeria, South Africa, Thailand and others have prioritised satellites to support social and economic development. The development of microsatellite technology and the role of private and commercial companies have contributed to making space increasingly accessible. The cost of launching a satellite into GEO (geostationary orbit) has decreased from an average of about 40,000 U.S. dollars/kg in 1990 to 26,000 U.S. dollars/kg in 2000. The cost is even lower for low-Earth orbit (LEO) as payloads could be placed in orbit for about 5000 U.S. dollars/kg in 2000. During the last decade, an average of 44 satellites was launched each year. Satellites dedicated to non-classified defense applications

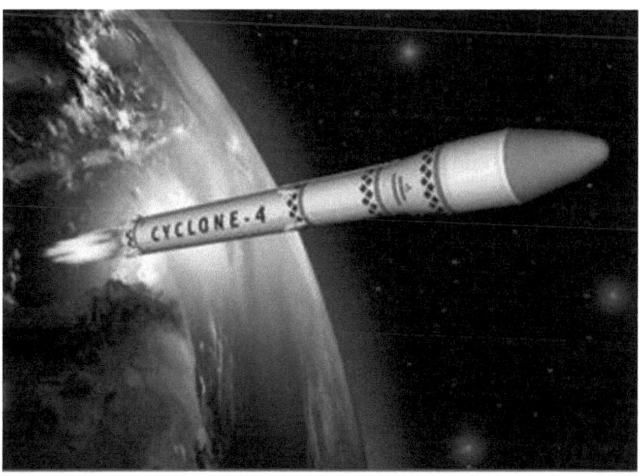

Fig. 1. *Brazilian-Ukrainian rocket Cyclone-4 (source: Alcantara Cyclone Space).*

represented 43% (19 satellites) while satellites for civil space applications were 57% (25 satellites). The increase in the number of actors gaining access to space has created greater inter-relation and complexity and has placed increased demand on available space resources. This situation will call for greater efforts to ensure coordination and implementation of international legal obligations (Figure 1).

6.1.2.2. Commercialisation

Space is generating a growing number of commercial activities mainly funded by private actors. In 2006, satellite services generated revenue estimated at 111.14 billion U.S. dollars. Initially limited to telecommunication services, individual consumers are becoming more important stakeholders in space as their demands are "expanding" to satellite navigation/positioning and remote sensing.

In the 11th edition of "Government Space Markets, World Prospects to 2017" Euconsult underlined the key trends, drivers and prospects for government space projects. According to this report while slower growth is expected for civil space programmes, defense-related funding should peak in 2012. More than 600 satellites are planned for launch in the next 10 years. In parallel, the regional breakdown will significantly evolve.

6.1.2.3. Development and security applications

Space technologies are considered as a major tool for development and security applications. The World Summit on Sustainable Development (WSSD) Plan of Implementation contains more than 10 specific references to Earth observation which clearly demonstrate the role that space technology can play in assisting sustainable development. Where space technologies are the most important component, the Global Earth Observation System of Systems (GEOSS) 10-Year Implementation Plan (endorsed by more than 70 countries) has as its aim to "qualitatively improve our understanding of the Earth system, markedly enhancing global policy and decision making abilities to promote the environment, human health, safety and welfare".

6.1.3. Factors to influence cooperation

The wide scope and benefits of space activity make it an ideal forum for international cooperation. This has been a common practice throughout the space age. The Outer Space Treaty (OST) of 1967 and the Declaration on International Cooperation in Exploration and Use of Outer Space adopted in 1996 established the basis and principles of this cooperation. Bilateral and multilateral initiatives have been established and range from assistance for capacity building to joint missions for space exploration.

Nevertheless, space cooperation is a result of a complex and unstable combination of several factors:

- Strategic goals of actors;
- Security and global issues;
- Number of actors gaining direct access to space;
- The importance of the private sector and the development of commercial activities.

Experience has demonstrated that developed regional political organisations create optimal conditions for cooperation. Two examples illustrate this point: the extension of the European Union accelerated the integration of several countries into the European space activities conducted under the umbrella of the European Space Agency as well as bilateral programmes, and the Sentinel Asia initiative has become an operational and cooperative programme for the use of space tools to enhance security and disaster mitigation in Asia.

6.1.4. The role of the United Nations

The United Nations through the Committee on Peaceful Uses of Outer Space (COPUOS) has historically concentrated on two main areas:

- The preparation of a legal framework for space activities at the global level.
- The promotion of space applications to encourage developing countries to access space benefits.

In 2007, COPUOS submitted to the United Nations General Assembly a draft resolution on space debris mitigation, largely based on guidelines developed by the Inter Agency Space Debris Committee (IADC). The adoption of this resolution was a first step in a process initiated by COPUOS to address the space security issue.

Among other critical issues that should be part of the COPUOS' agenda and should be discussed as they will impact the fair and responsible use of space are

Fig. 2. *ESA Don Quichotte mission to study NEOs (source: ESA).*

space traffic management, environment protection on the surface of the moon and other planets, protection against near-Earth objects (NEO), etc. (Figure 2).

6.1.5. The way forward

Only through cooperation can the full potential of space technology for economic and social development be guaranteed with due respect for the relevant interests of all countries. Indeed, unilateral actions in outer space are no more tenable than unilateral actions in any other area of global concern (e.g., security, climate change, natural disasters and environmental degradation). Nations have become more integrated and such integration demands cooperation to strengthen collective security.

But with regard to the diversity of the players' needs, strategies and levels of development, the cooperation mechanisms should be diversified and rely on key elements.

6.1.5.1. Political commitment

Although it may sound obvious, it is important to bear in mind that cooperative actions cannot be productive unless they are based on the strong political commitment of all the participating players. Two examples illustrate this principle. The GOE process was originally initiated by, and continues to benefit from, the strong commitment of political players that made it possible in a short period to work out an ambitious and integrated action plan (GEOSS). More recently, on December 2007 in Lisbon, the European Union and the African Union endorsed a Partnership on the Information Society, Science and Spaces Technologies (I3 S). As an operational result of this partnership, the GMES programme was extended to Africa (GMES-Africa).

6.1.5.2. Transparency and confidence building

As is argued by Rihanna Tyson, Transparency and confidence-building measures are key elements "to promote dialogue and interaction, to facilitate information sharing and increase trust between States". Such measures depend on access to data and information to be effective. Two categories of data can be identified: scientific data which promotes the sharing of knowledge between researchers in order to facilitate scientific and technical progress and environmental data that enhances capacities for planetary environment protection and the prevention of major risks as well as the management of natural disasters.

6.1.5.3. Collaboration mechanisms

The commitment of current and future space actors to achieve global engagement in space security also depends on collaboration mechanisms. These should be based on:

- A clear vision expressed and promoted by a coalition of governments.
- An ongoing planning and coordination forum to establish common goals and objectives and to play a facilitating role in the development of mutually beneficial programmes.

These collaborative mechanisms should also build on existing constellations of intergovernmental structures. More structures would create confusion, conflicts and "dispersion" of resources and efforts.

6.1.5.4. Ethics of space

As a shared international concern, at the level of the global community, the ethics of space must adopt a strategic approach to space policy-related decision-making especially where it concerns the security aspects of space and its uses. To ensure fair and responsible use, the basis and principles that should guide the ethics of space are as follows:

- Benefit sharing (communication, disaster and crisis management, environment management and protection etc.) as well as risk sharing (use of nuclear energy sources, space debris, electronic surveillance, etc.).
- Resource sharing through an equitable distribution of resources especially limited resources (geostationary orbit, spectrum, etc.).

However, the international community has to be more vigilant as the current trend seems to be to replace the original political logic with a commercial one. The space community should prevent economic dogmatism from depriving some populations and countries of space benefits.

6.1.5.5. Improvement of the institutional/legal framework

Cooperation in space must be anchored in a legal framework as a prerequisite for cooperation in addressing universal threats and security issues (climate change, food security, social instability, etc.). The current legal framework is based on the Cold-War era and focuses largely on regulating the States' military affairs. Worldwide consultations should be launched with a view to reaching a common understanding of rules and norms and to limit divergences and conflicts resulting from competing interpretations. With regard to the evolution of space activities and the respective changes in space actors, the space community should urgently start a process to reinforce and up-date current international treaties and conventions and to "fill in" the gaps in existing space law. The last mandatory measure was adopted 30 years ago in 1979.

The reinforcement of the international legal framework within the United Nations system is becoming more urgent as we are increasingly witnessing a proliferation of guidelines, codes of conduct, voluntary measures, etc., that are worked out by non-governmental organisations without taking into account the needs, views and positions of numerous users and nations.

6.1.6. Conclusion

To commit current and future space actors to a fair and responsible use of space is a vital task that should be undertaken under the aegis of the United Nations and in compliance with its principles, in particular the Treaty on Principles Governing the Activities of States in the Exploration and Use of Outer Space, known as the Outer Space Treaty (OST). The role of governments and States is central to ensure the fair and responsible use of space and to produce control mechanisms and systems. Legal instruments should be updated and reinforced to serve as the framework for the fair and responsible use of space.[156]

[156] The following references were used for this article: Pompidou, Alain. L'Ethique de la Politique Spatiale. Paris: Commission Mondiale d'Ethique des Connaissances Scientifiques et des Technologies, UNESCO, 2000; Wolter, Detlev. Common Security in Outer Space and International Law. Geneva: UNIDIR, 2006; Euroconsult, Government Space Markets, World Prospects to 2017, 2008; Tyson, Rhianna. "Advancing a Cooperative Security Regime in Outer Space" May 2007. Global Security Institute Policy Brief. 15 June 2009. http://www.gsinstitute.org/gsi/pubs/05_07_space_brief.pdf; Space Security Index. 15 June 2009. http://spacesecurity.org/United Nations. Promotion of International Cooperation. Background Paper for the Third United Nations Conference on the Exploration and Peaceful Uses of Outer Space. UN Doc. A/CONF. 184/BP/12 of 8 May 1998. Vienna: United Nations.

6.2 The role of international organisations for the fair and responsible use of space

Kazuto Suzuki

6.2.1. Introduction

International society is composed of sovereign States. The membership of international organisations such as UNCOPUOS is limited to sovereign Member States and they are the ones who make the final decisions. If there is no agreement or consensus among sovereign States, there will be no regulations. However, we must ask whether the players in space are limited to only State entities. Many commercial organisations are already active in satellite communications and Earth observation and may become active in human space flights in the near future. We cannot simply assume that States are the only actors in space.

Moreover, the assumption that legally binding rules in international law always modify the behaviour of States should be scrutinised. There are various international rules and regulations but the degree to which States comply with them is quite diverse, particularly when there are many "new" States entering into the world of space activities, such as Iran, North Korea and so on. These States may not share the same idea and normative understanding. It is misleading to believe that once a rule is set, it will automatically modify the behaviour of States and settle international disputes.

On the other hand, it is also not always true that States take actions to maximise their own interests in complete disregard of rules and regulations. There have been suspicions that the United States action towards Iraq was not legally appropriate, but what is important is that even the United States under the Bush Administration sought to find and establish the legality of its action. Although its action was unilateral and did not respect the international order, it tried to convince the international community that its action was legal. Moreover, there are many States that take actions even when those actions are not consistent with their national interests.

6.2.1.1. International organisation as the core of "global governance"

The three assumptions underlying theories of international relations as discussed above tell us that the means by which international order is set up are becoming more complex and difficult. It is not as simple as establishing a new treaty or rules, or letting powerful States establish their own order for the fair and responsible use of space that would actually be in their favour. In this context, the concept of "global governance" could be useful for considering the future of the fair and responsible use of space. Global governance can be defined as "collective efforts to identify, understand, or address worldwide problems that go beyond the capacity of individual States to solve".[157] This concept is designed to explain how international order can be achieved in the absence of global government.[158] Global governance would not provide the final solution for the problems of fair and responsible use of space, but it suggests that if States and non-state actors can collectively work on identifying, understanding and addressing the problem of establishing a norm for fair and responsible use of space, there would be a chance for us to see safe and stable conditions in orbit and on celestial bodies.

Finding collective identity and understanding is especially important in recent years because of the development of new services and the increase of space-faring nations. New services, such as private suborbital manned flight or privately developed commercial launching services, require new sets of rules and regulations for safe and orderly service provision. Furthermore, the emergence of new actors, such as Iran and North Korea, which have little experience in space activities and frequently not consistent with global norms in many respects, poses other problems. The objectives of their space activities are to establish national technological capabilities and to increase their international political importance. In other words, the fair and responsible use of space is not the first priority for them if it obstructs their goals. Thus the concept of global governance, despite its difficulties, is of the utmost importance today.

To establish a solid and satisfactory global governance system it is imperative that international organisations play a role as a "core" of global governance. Although an international organisation is not a world government or an entity that stands above sovereign States, it would provide a forum for exchanging, deliberating on and converging the interests and visions of Member States to establish a common understanding, action and identity for the fair and responsible use of space. Also, once an international norm is accepted by the Member States of the international organisation, it would have significant impact on private and non-state actors in space.

The reason for stressing international organisation is largely because it would be difficult for only a small number of actors to set up an international norm for the fair and responsible use of space. It may be possible to achieve the goal through a unilateral top-down approach in which one State, such as the United States, or a small number of space-faring States, imposes a rule for global space regime. However, the success of an international norm depends on the willingness of all participants, including those who have the potential to participate in the use of space, to abide by it. At present, the only possible locus for including all current and potential participants would be an international organisation.

6.2.2. Power resources of international organisations

International organisations are often regarded as a watchdog without sharp teeth. Unlike national governments, they do not possess coercive forces to ensure that Member States comply with international law and rules. However, international organisations have certain kinds of power for persuading States to behave in accordance with their decisions.

Fig. 3. *COPUOS (source: Secure World Foundation).*

The first kind of power is authority. No one denies that international organisations are the only forums for representing the entire international society, reflecting international concerns, discussing important issues and taking decisions. The quasi-democratic nature of an international organisation (i.e., each State has one vote but individuals have no direct representation) further strengthens the authority of an international organisation. In the case of UNCOPUOS, most decisions are taken by consensus, which means that even a small country without space capability would be able to express its opinion about the decision and sometimes it can block the decision-making process. At the same time, a decision taken by consensus inevitably puts moral obligations on States to comply with the decision. In the theory of international law, agreement binds the action of States. Thus, consensus decision-making binds the behaviour of States since they have implicitly agreed to be bound by the decision and have not expressed their opposition to it. The authority and democratic nature of international organisations are their strengths (Figure 3).

Additionally, an international organisation has the power to establish the agenda and define what is important to the international community. If there are conflicting visions of the fair and responsible use of space, an international organisation would be able to set the agenda in favour of one side to the exclusion of other issues. Furthermore, the issues that are taken up and discussed in the international organisation would be covered by international newspapers and news reports, so that the agenda and decision will be regarded as a central concern for the international community. The discourses and general understanding of the issue would be a focus of international discussion and would gradually lead towards greater coherence and conversion of the views of Member States and the world public in general. Good examples of this are the anti-whaling movement and the treaty on banning landmines. It is mistake to underestimate the power that international organisations have in agenda setting and norm-creation.

Moreover, international organisations do have some teeth in terms of imposing sanctions on States that break rules and decisions. An international organisation can make a decision recommending and requesting Member States to take action against rule breakers and ultimately it can appeal to the UN Security Council which is capable of taking legally binding decisions for coercive actions. Also, international organisations are able to impose certain sanctions within the organisation such as suspending membership. Its teeth may not be sharp, but still an international organisation does have teeth with which to bite.

6.2.3. Limitations of international organisations

Although international organisations have certain power resources to alter the behaviour of States, there are limits to what they can do. First, the question of membership can be of concern. Based on the principle of "agreement binds", it would be difficult to convince States to comply with decisions taken by an international organisation if the State is not a member of the organisation or does not participate in its decision-making. A good example is the case of India and Pakistan and the Non-Proliferation Treaty (NPT). They did not join the NPT regime because they did not want to be constrained by it, due to the unequal nature of the regime. Although UNCOPUOS has 69 members most of whom have specific interests in space activities, it would be difficult to expect some non-members such as North Korea to comply with the decision of UNCOPUOS.

However, while it would be better if the international organisation consists of many States with space capabilities, the more the membership expands the more difficult it is to reach consensus. Because States have a wide variety of specific interests and objectives in their space activities, it could be difficult to set up common ground or understanding that satisfies all participants. The most capable and autonomous States such as the United States, Russia and China may not be happy if there are too many restrictions on what they would like to do, whereas developing space-faring States could try to avoid additional costs in developing their spacecrafts by using lower level safety or debris-related features. The difference of interest comes not only from the different developmental stage of States but also from the strategic objectives of the space activities. Some States do not want to disclose technical and technological information about their spacecraft if it is related to security measures. These differences of interests and objectives are difficult to reconcile even with the authority of an international organisation.

Furthermore, even if there was consensus among Member States of international organisations, it would be difficult to impose strong measures for member compliance because agreement in international organisations is often recommendatory rather than legally binding. Since the failure to achieve ratification of the Moon Agreement, most international rules have taken the form of guidelines, which have weaker binding power than treaties or agreements.

In addition to these difficulties and limitations, it would be difficult for an international organisation to act beyond its mandate. For example, the mandate of UNCOPUOS is "to review the scope of international cooperation in peaceful uses of outer space, to devise programmes in this field to be undertaken under United Nations auspices, to encourage continued research and the dissemination of information on outer space matters, and to study legal problems arising from the exploration of outer space"[159] In other words, UNCOPUOS cannot determine what States can and cannot do in space but only reviews, encourages and studies the issues. Of course, there are cases where the mandate of an international organisation has been extended by the agreement or reinterpretation of the mandate among Member States (such as the case of the UN Security Council). This could also happen to UNCOPUOS but currently there is no explicit agreement among Member States to do this.

6.2.4. How to overcome limitations

6.2.4.1. Leadership in international organisations

Limitations, not only for UNCOPUOS but for all international organisations, must be overcome if international organisations are to be able to play a full role in the fair and responsible use of space. How can this be done? With regards to strength and authority there are several ways of enhancing the capability of international organisations, the first of which is the exercise of leadership. Although interests and objectives differ between Member States, it would be possible to change their minds and convince them by using scientific and moral discourses with strong leadership that is committed to achieving consensus. Scientific discourses would be particularly effective because they are difficult to challenge. For example, the Intergovernmental Panel on Climate Change (IPCC) has played a huge role in convincing many States to accept the necessary measures for reducing global warming gases. Moral discourses are also a very strong weapon for appealing to civil society and changing the attitudes of sceptical States. The International Campaign to Ban Landmines (ICBL) successfully employed such tactics to link human rights and security issues by exposing the inhumanity of landmines. Canada, as host of the Ottawa Process, exercised its leadership jointly with the ICBL by providing various diplomatic services and convening informal discussions while

Fig. 4. *ICBL logo (source: ICBL).*

ICBL focused on garnering public support through the media and civil society (Figure 4).

The power of discourse is not the only power that leaders can exercise to achieve consensus. True leadership does not come from merely holding a high-level institutional position in the structure of an international organisation. Leaders must find their own ways of strengthening their position. The most common requirement is conviction. Leaders use the scientific and moral discourse to convince other Member States, thus demonstrating their commitment and values.

But some leaders may exercise power through intimidation by threatening negative consequences if other Member States do not follow the leader. For example, the leader may threaten to reduce international aid or to withhold the provision of assistance if other Member States do not comply with its leadership. Similarly, leaders can exercise the power of reward or positive sanction on other Member States. This could lead to positive results if other Member States are hesitant or unwilling to follow the leader. In both these cases, leaders have to possess certain resources to enable them to use positive and negative sanctions – threat or reward – on other Member States.

If the leader does not have such resources to leverage other Member States, the leader can still exercise power by being a moral authority or providing a role model for other Member States. The leader can unilaterally demonstrate its commitment and values by sacrificing its own interests and appealing to other Member States to follow this example. For example Nordic States, such as Sweden and Norway, as well as Canada have been leaders in UN peacekeeping operations by demonstrating their commitment to the concept of peacekeeping by the United Nations through their constant contributions to UN peacekeeping forces. This unilateral action attracts respect from other countries and has become a model for such activities.

6.2.4.2. Transfer of norms

The other aspect of enhancing the capability of international organisations in the fair and responsible use of space is the transfer of norms. As in the case of the anti-landmine movement where the norms of human rights issues were transferred to the sphere of security issues, international organisations have the possibility of introducing a norm that was developed elsewhere. In the case of space, there it is possible that the norms developed in the United Nations Convention of the Law of the Sea and the Antarctica Treaty can be applied to the fair and responsible use of space. These treaties and agreements are products of the necessity of having a common ground for all States to be able to preserve areas where there is no jurisdiction. Space, as a new frontier of human activities, can be regarded in a similar context. Lessons can be learned from the Law of the Sea and the Antarctica Treaty for preserving outer space as a common sphere for all mankind and establishing rules and regulations for the fair and responsible use of space for all States and actors.

To some extent this relates to the question of the limitations of international organisations as discussed above. UNCOPUOS deals exclusively with space matters without linking this to other areas of activity. It is my opinion that there should be much more interaction with other forums in the UN system for transferring the norms and lessons that have been learned in other domains. It is also important to mention that non-governmental actors, such as in the case of transferring norms from anti-landmine activities to anti-cluster bomb activities, should play a significant role in linking precedents from the Law of the Sea or Antarctica Treaty to space laws.

6.2.4.3. Moral obligations

As discussed, space can be regarded as a common ground for all mankind since no one can territorially occupy the physical space in orbit (other than celestial bodies) and anything that happens in the Earth orbit can influence everyone else using the orbit. This applies regardless of whether the State is an "advanced" or "developing" space-faring nation or whether the operator is civil, commercial or military. It is the law of physics that dictates the movement and activities of spacecraft, manned or unmanned. If there is no stable and safe order in space, it is to the loss of everyone who has a stake in space. Of course, "advanced" space-faring States that have large numbers of spacecraft in orbit and are more interested in using space could be affected by space debris and satellite collisions while "developing" States also have concerns since they are not sufficiently equipped with measures to avoid and mitigate debris and collisions. A stable and safe order in space is crucial for commercial operators because of the risk and cost of their operations but it is also important for the users of those services who might suffer damage if there is no stable and safe order in space.

However, some space-faring States, particularly the United States and China, have regarded space as a new battlefield where one State can dominate and exclude others. The concept of "space power" has been commonly discussed in the academic field with the understanding that space capability strengthens the strategic power of a State.[160] The destruction of the weather satellite by China in 2007 created over 2000 known debriss and significantly endangered other spacecrafts. Similarly in 2008, the United States shot down one of its own military reconnaissance satellites that was out of control. Although the U.S. action was closer to the atmosphere, it still did not provide a 100% guarantee that no debris would be created. These actions are representative of the view that space is a place that powerful States can control and from which they can exclude the entry of other States. Nevertheless, it is almost impossible for any State to control the Earth orbit and to limit other States from becoming active in space. It would be utterly unwise to threaten other States by destroying spacecraft in orbit because it would create debris and as a result might harm its own spacecraft.

Thus all space-faring States, whether "advanced" or "developing", whether civil, commercial or military operators, have a moral obligation to use space fairly and responsibly because it is also in the best interests of all space-faring States. An international organisation, therefore, would be in a position to exercise its power and authority to convince all Member States to recognise their moral obligations and adhere it to its decisions.

6.2.5. Towards establishing fair and responsible global governance

6.2.5.1. Norm entrepreneur

Establishing global governance for the fair and responsible use of space would require a "norm entrepreneur": i.e., someone who takes the initiative and exercises leadership in creating and widening the discourse. For example, Henri Dunant was the norm entrepreneur for establishing the International Red Cross, Dag Hammarskjöld was the norm entrepreneur for establishing UN peacekeeping operations, the IPCC was the norm entrepreneur for the Kyoto Protocol and the United Kingdom was the norm entrepreneur for abolishing slavery. It is not clear yet who or which State should be the norm entrepreneur for the fair and responsible use of space but it is our expectation that the European Space Policy Institute and the Secure World Foundation, who have already been active in this field, will develop a norm and attract the attention of the media and a number of governments.

As we discussed above, however, the norm entrepreneur cannot lead negotiations at the intergovernmental level without an alliance with enthusiastic States interested in international organisation of this area. It is imperative that the messages of these norm entrepreneurs be heard by some States and that alliances are formed with them. Dunant was not able to achieve his goal without forming an alliance with Switzerland; Hammarskjöld could not have been successful without Canadian Foreign Minister Pierson; and the IPCC would not have been a Nobel laureate without allying with European States. Thus, it is important that norm entrepreneurs for the fair and responsible use of space find credible partners at the intergovernmental negotiations level for promoting their norm. In this regard, the Code of Conduct that was proposed by the European Union under the French Presidency would be a great step forward for establishing an alliance with norm entrepreneurs such as ESPI and SWF to promote their values at both intergovernmental and civil society levels.

6.2.5.2. Mutuality

When we discuss the fair and responsible use of space, we need to be careful about the meaning of "fair" and "responsible". Fairness comes only when the rules are applied equally to everyone without discrimination. When norm entrepreneurs and allied States promote a new norm and understanding of the use of space, it should not be unilaterally imposed on other Member States. Also, some advanced space-faring States should accept the concept of mutual responsibility without reservations. Capable States are often not willing to be constrained by international obligations because they want to preserve their freedom of action as much as possible. However, if there were exceptions to the rules, it would be difficult to convince other participating Member States to comply with those rules. As discussed above, capable States are the largest beneficiaries of the fair and responsible use of space because they are the ones who are most exposed to the risk of colliding with space debris and other dangers in orbit.

6.2.6. Conclusion

The fair and responsible use of space is not an ideal or fantasy. Rather, it is a necessity for every space actor in the world. A stable and safe space environment is an absolute requirement for civil, commercial and military space actors. Without a stable and safe environment, it would be difficult for any space actors to provide continuous services that are vital for activities on the ground. Many, if not all, States are significantly dependent on space assets for providing commercial, social, political and military services. The security of those assets is the number one priority for mankind.

The space environment is already contaminated to a certain degree, particularly with the destruction of the Chinese satellite in 2007 and the 2009 collision of the Iridium and Cosmos satellites. The increasing risk of space debris is threatening the future sustainable use of space. Currently, the only possible way to avoid collisions is by strengthening the debris monitoring system through Space Situational Awareness (SSA) but the manoeuvring satellites would be too costly if there were too much debris and too many spacecraft. Thus it is imperative for all space actors to agree upon international rules for the fair and responsible use of space and the only way this can be done is through an international organisation such as UNCOPUOS.

After all, the only law that dictates space activity is the law of physics. Man-made law and rules can be broken by malicious behaviour but no one can cheat the law of physics. The law of physics is also the fairest law because it equally distributes risks and consequences. No matter whether the operators are civil, commercial or military, the consequences of unfair and irresponsible use of space would be fairly distributed to everyone. Whatever the United States, China or any other space-faring States try to achieve, they cannot escape from the law of physics and have to accept the consequences of their actions. In other words, there is no choice but to use space fairly and responsibly.

[157] Weiss, Thomas G. "Governance, Good Governance and Global Governance: Conceptual and Actual Challenges." Third World Quarterly 21.5 (2000): 795–814.
[158] Rosenau, James and Ernst-Otto Czempiel, eds. Governance Without Government: Order and Change in World Politics. Cambridge: Cambridge University Press, 1992.
[159] United Nations General Assembly. International Co-operation in the Peaceful Uses of Outer Space. Resolution Adopted by the General Assembly. UN Doc. 1472 (XIV) of 12 Dec. 1959. New York: United Nations.
[160] France, Martin E.B. "Back to the Future: Space Power Theory and A.T. Mahan", Space Policy 16.4 (2000): 237–241.

6.3 Youth contributions to the debate on space security

Agnieszka Lukaszczyk and Alex Karl

6.3.1. Youth involvement in space – is there a voice?

It is often assumed that space is an "Old Boys Club", very exclusive and difficult to penetrate for a young person. Ironically, people advocate that more outreach must be done because the sector is suffering from not having enough capable and competent people to take over jobs of those who are soon to retire; yet, there is no easy access for young space enthusiasts to join the sector and to become its legitimate members.

Clearly many young people around the world are fascinated by outer space. Most boys and girls at some point have dreamt to become astronauts and to explore the unknown. What is special about the space sector is that people coming from a variety of backgrounds can contribute. It is not only about the technical aspects of the field but also about policy, legal issues, societal questions and more.

Space security is one of the topics that is considered to be "hot" by the younger crowd. Issues of near-Earth objects, space debris, space traffic management or the danger of an arms race in space are fascinating and attract the attention of young space enthusiasts. There are several venues through which a young person can contribute to the debate on space security. One such platform is the Space Generation Advisory Council (SGAC), which facilitates projects, events and discussions on space security among other things (Figure 5).

6.3.2. What is the space generation advisory council (SGAC)?

The Space Generation Advisory Council in Support of the United Nations Programme on Space Applications (SGAC) aims to represent young people (students and young professionals) between the ages 19 and 35 to the United Nations, other international organisations, governments and space agencies.

Fig. 5. *The SGAC logo (source: SGAC).*

SGAC was created in 1999 at the UNISPACE III Conference where space leaders from all over the world came to a conclusion that youth should have a voice at the United Nations. Therefore parallel to UNISPACE III, the Space Generation Forum took place where 160 young people from 60 countries worked together and produced and adopted the Declaration of the Space Generation. In order to fulfil its mandate, SGAC has been given permanent observer status to the UN Committee on the Peaceful Uses of Outer Space (COPUOS) and consultative

Fig. 6. *SGAC representative addressing COPUOS.*

status with the UN Economic and Social Council (ECOSOC) so that it can regularly report its activities to the UN (Figure 6).

SGAC has a true global network with over 4000 members worldwide who communicate and work together via e-mail lists, online forums and projects, local, regional and international workshops and conferences as well as various grassroot activities. The organisation has two Regional Coordinators (RCs) from the six UN regions, and currently 66 National Points of Contact (NPoC). SGAC's focus is to impact policymaking and thus advise political decision makers based on the opinions and ideas of the youth of the world. The policy input includes regular reports to the Member States of UNCOPUOS, including the UNISPACE III Action Teams. In addition, SGAC gives input to national and international space agencies as well as governmental entities dealing with space. For instance SGAC was the only NGO invited to the Space Summit in 2002 in Houston, which gathered major space companies, high-level space experts and heads of space agencies to discuss the future of space. More than that, SGAC was invited by the European Commission to provide the perspective for the future European space workforce during the consultation process for the Green Paper on the European Space Policy.[161]

Having such impact on policymaking motivates the members of SGAC to participate in its activities on national, regional and international levels. Because members know that through SGAC they have access to the UN and various high-level space persons, they can address the issues they care about and have their voices heard. For instance, in 2007 the members of the SGAC South American region wrote reports on the space situation in their respective countries, which were then delivered to the COPUOS delegates.

In addition to policy focus, SGAC carries out a number of projects around the world including projects on education, space security and outreach programmes. SGAC sends its members to local schools to teach astrophysics, astrobiology, astronomy, science and technology, sustainable development, etc. It also organises space awareness days, movie nights, workshops, conferences, space parties (Yuri's Night), and technical activities. Hands-on activities such as the "Under the African Skies" project help to build knowledge and confidence but, most importantly, they allow SGAC to teach people in developing countries about the impact of space technologies in their day-to-day lives and how they could benefit even further from a number of space applications.

Because of its good relationships with the UN Office for Outer Space Affairs (OOSA) and UNESCO, SGAC is able to send some of its members from developing countries, fully funded, to a number of international conferences and workshops. This gives those young people the opportunity to interact and learn from experts in the field. But most of all, it allows them to present their ideas and thoughts in international forums where they can gather advice and input from people they would normally have no access to.

In summary, due to its international mandate, SGAC is able to carry out many useful activities around the globe. Its network has grown significantly in the past 10 years, thus it likes to call itself a "network of networks" as it not only represents its members on the international level but it also works as an umbrella for other student and youth groups around the world.

6.3.3. The visions of the space generation advisory council

During the foundation of the Space Generation Advisory Council, the Declaration of the Space Generation was written to express the visions of youth on space. The students and young professionals participating in this very important project felt very strongly about certain aspects concerning the utilisation of space as they were very aware that they are the future players in the space sector and thus their voice should be heard:

> "We, the Space Generation, representing the worldwide visions of youth, commit ourselves to ensure the future of humankind."
>
> "In leaving the Earth's cradle, in the quest for understanding our place in the Universe, we are entrusted by the next generations with the sustainable development of the planet for our peaceful future."
>
> "We, the Space Generation, regardless of culture, language and creed must ensure that space exploration will improve the quality of life for the benefit of all humankind."
>
> "We express the hope and the conviction that our common future ought to proceed *ethically*, with an *understanding* of the long-term consequences of our actions, and with *all of humanity* walking forward together as one."[162]

Furthermore, in 2003 SGAC stated the following aim in its strategy document: "Advancing human development through the peaceful uses of outer space". Since then SGAC has continued to demonstrate that the peaceful uses of outer space are of high importance to the members of the organisation.

Since 2007 SGAC has been actively collecting contributions towards the formulation of multidisciplinary visions of the youth for the next 50 years of space activities. SGAC members coming from all kinds of backgrounds and regions agreed on the following three themes:

1. Ensuring the Survival Interests of Humanity.
2. Space for the Benefit of all Humanity and of our Environment.
3. Advancing the Frontiers of Science and Technology.

In particular, the recommendations of Theme 1 called for space governance mechanisms such as international space law that would help ensure the peaceful and sustainable use of space.

The following examples pertinent to space security emerged from the study:

Space governance: Space efforts need to be led by all nations collectively including space-faring countries under the umbrella of international organisations such as the United Nations. Apart from some very important revisions to the Outer Space Treaty (OST), there is a need to include the registration of defence-related missions in order to reduce the secrecy and security threats that still surround business within the space environment. It was also noted that important revisions to the OST are needed, particularly revisions pertaining to governance of the destinations of near-term exploration such as cis-lunar space, the Moon surface and Mars. There was strong support for internationally cooperative efforts for space exploration and to help humanity benefit from the coordination of research efforts from all countries.

Peaceful uses of outer space: It was very apparent in the survey that it is of crucial importance to the respondents to make outer space secure and used only for peaceful purposes. They recommended that space agencies and governments continue their dialogue on the future military uses of space in order to establish rules that would benefit society in general. When thinking about outer space, young people no longer see each other as citizens of individual countries but as citizens of the world. Rather than competition among nations, they favour cooperation at the international level with a deep understanding of the necessity for global collaboration in outer space. In addition, the respondents felt strongly about the Disaster Monitoring Constellation in the sense that it is a globally coordinated, cooperative effort to protect Earth. They would like to see an international effort to create a "global network of observation" to encourage the free flow of information to minimise the potential threat of global disasters and to redouble efforts to observe all large near-Earth objects (NEOs) by 2010. They encouraged an increase in public awareness regarding the potential for current and developing space technologies created for the purpose of averting potential disasters originating from Earth or from the threat of NEOs.

The role of developing countries in space exploration: As mentioned above, outer space can serve as a unique platform for the citizens of the world to collaborate in various endeavours. Hence, the benefits of space technologies should be accessible to all. Today, we are witnessing a resurgence of interest in lunar exploration. Worldwide, the motivation of space agencies to spend more time and money on lunar research activities is increasing. Improving the practicality and reliability of advanced technologies in space transportation is crucial for the enduring exploration of space. It is important to recognise the economic opportunity that has inspired previous exploration ventures and remember this for future exploration strategies. Reports and studies such as the 2007 International Lunar Decade by the Planetary Society promote international collaboration to contribute to technological progress. This cooperation can be expanded to encourage developing countries to become involved in space programmes. Respondents encouraged fostering capacity-building between countries, intergovernmental organisations and/or NGOs. They proposed achieving this by advancing space capabilities within developing countries. They encouraged developing nations to pool resources and become involved in space-related operations by promoting education and research and creating an infrastructure to permit the development of a potential space programme. Awareness and information on space exploration must be encouraged in developing nations.

Space education and outreach: There is a lack of space education in schools, especially in developing countries. Students are not aware of the opportunities that exist in fields such as remote sensing applications and satellite communication and are not aware of future space programmes in their country. Through more comprehensive space education, they will be able to apply this knowledge to solve specific problems within their community, thus contributing to the economic development of their nation. Creation of a Global Space Education Curriculum – expanding space education in schools on an international level by convincing governments and schools to include space curriculum in classrooms – has been recommended. These programmes will raise space awareness as well as stimulate student interest in studying science and engineering. Organisations such as UNESCO, among others, should play an active role in encouraging educational programmes in space research. Space agencies should regularly inform the public, especially people from developing countries, of the benefits of space technologies by setting up specific workshops and educational events. Young people pointed out that it appears as if many agencies, organisations and governments are aware of the poor information distribution when it comes to space topics; however, very little progress has been achieved in this area.

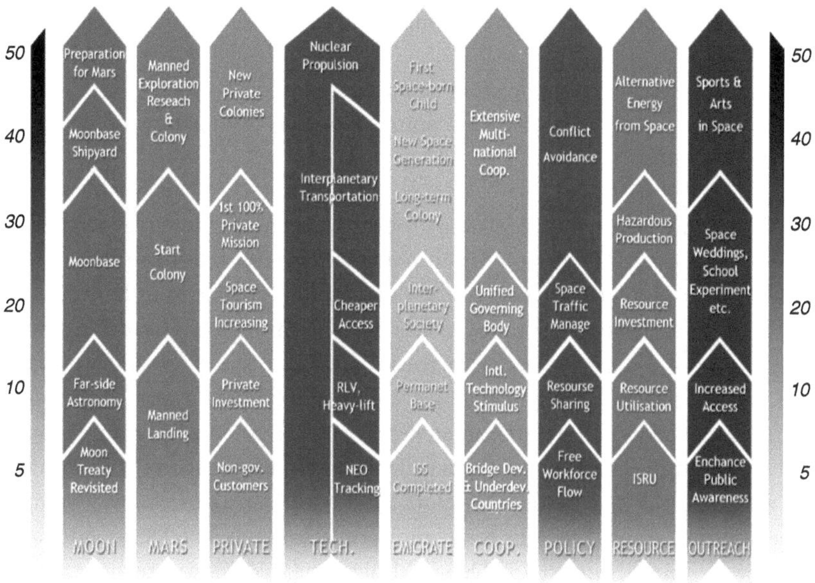

SGAC 50-Year vision roadmap

Fig. 7. *Vision roadmap (source: SGAC).*

These visions were drawn into a roadmap (Figure 7).

The vertical axis shows the number of years into the future while the horizontal axis is divided into different categories such as Moon, Mars, Private, etc. The blue arrow to the right depicts the Policy dimension of the 50-year roadmap as our members see it. Free Workforce Flow basically addresses the wish to encourage exchanges in expertise and knowledge. This is followed by Resource Sharing. Resources in this context are space itself and its resources. Space Traffic Management and Conflict Avoidance conclude the outlook.

6.3.4. Safeguarding space

To sum up, SGAC visions and recommendations have consistently pushed for developing space in a way that safeguards space for all of humanity. Safeguarding space, for the purpose of this paper, means to ensure the long-term viability of all humanity to use space for peaceful purposes. The consequences of that are that space should be kept free from any activities that:

1. are against the spirit of the peaceful purposes enshrined in the Outer Space Treaty;
2. inhibit the use of space by other actors;
3. in any other way destroy the finite resources or usability of the space environment.

When contemplating space security for the future generation in general and in particular through the 50-year vision roadmap, SGAC has identified four key issues that need to be addressed in the short term to provide long-term security.

The first issue is space debris. SGAC is very concerned about the rising amount of space debris as it represents an increasing hazard to spacecraft and astronauts. The greatest threats are fragmentation events due to exploding spent rocket stages or debris caused by collisions between debris as these potentially create exponentially more debris. In the long run this will inhibit sustainable access to space and its use as a resource for all parties to explore and to utilise in a peaceful way for many generations to come. Essential for a long-term solution is an effective mitigation strategy. Within UNCOPUOS Debris Mitigation Guidelines were suggested and adopted by Member States. However these guidelines are voluntary and not legally binding. Closely linked to the mitigation of space debris is surveillance capability. A greater resolution is needed to track objects that have the potential to disable a spacecraft, namely all objects larger than 1 cm, while currently only objects larger than 10 cm can be tracked in LEO. Greater resolution would be necessary to complement the mitigation guidelines to provide a long-term solution for space debris. Further, international cooperation to share relevant data would be beneficial to all actors involved.

Space Traffic Management is a logical step from space debris mitigation as it would also enable a comprehensive collision avoidance structure. While currently not much can be done about debris–debris collisions, active spacecrafts are able to do collision avoidance manoeuvres with just small amounts of fuel if an impending impact is predicted in time. With ever-increasing numbers of space actors as well as objects in Earth orbit, space traffic management is a logical necessity not only to avoid collisions but also to guarantee unimpeded and more efficient use of space resources for all humanity. The question here is not *if* but *when* it will be implemented. It must be clear to all space actors that having a space traffic management system in place will contribute to reducing the loss of working satellites, sustaining the use of space resources and providing an asset to the security of space for the coming years.

While not immediately related to space security in the short term, further clarifications within the space law regime are needed to resolve issues that will arise in the long-term as there is a general lack of basic laws of space conduct. In the future there might be conflict over land and resources off Earth, not necessarily solely among States but possibly involving commercial and private companies as well. The young generation strongly supports initiatives that aim at an accelerated development of space technologies within the private and commercial sector for the peaceful uses of outer space for all humanity. However, initiatives like this will eventually lead to a larger number of non-state space actors. Resulting from that, even if technology is invented in a peaceful environment for peaceful purposes, it cannot be ruled out (and it appears rather likely) that certain issues related to security might arise in an environment that is not sufficiently legally covered. This possibility was not perceived 40 years ago when the first Space Treaties were drafted and it needs to be addressed in a civilised manner within international space law to ensure sustainable access to space and utilisation of it and its resources for all humanity. It would also serve as an example for other areas of rapid new advances.

The final key point is conflict avoidance. Space weapons, due to their negative influence on the space security situation, and aggressive acts such as Anti-Satellite (ASAT) activities, that create large amounts of space debris, should be prohibited. As mentioned earlier, space debris is a serious concern because it might prohibit future generations from accessing and utilising space in a sustainable manner. Kinetic ASATs are therefore threatening everything our generation wants to do in the space environment. As the Executive Council of the Space Generation Advisory Council observed through the discussions happening on our member email discussion list and online forums, the ASAT tests that were conducted within the past 15 months triggered an intense reaction amongst the young generation who were and still are very concerned about these developments. It even led to the establishment of a working group by concerned youth and continues to be discussed and carefully analysed by them.

6.3.5. Conclusions

Deriving from the above-mentioned points, SGAC recommends the following:

1. Create a treaty through UNCOPUOS to make space debris mitigation legally binding and increase resolution of surveillance capabilities and encourage the sharing of relevant data.
2. Initiate a working group on space traffic management as well as address a framework for rules of the road possibly through an Inter-Agency Space Debris Coordination Committee (IADC) Working Group and then further to COPUOS.

3. Avoid conflict – prohibit weapons and aggressive acts from space.
4. Address several open issues regarding space governance (lunar governance and property rights).

The SGAC studies illustrate the views and opinions of young people around the world who are very much concerned about space issues. Just because these studies are done by students and young professionals they certainly should not be taken lightly. On the contrary, they should be taken as fresh and honest contributions towards building up space policy. Due to the lack of political pressures, young people tend to look at things somewhat differently from those in power, thus offering a sincere, concerned and unbiased view concerning the future here on Earth and in outer space. Their pure dedication towards the case for space, and hence, space-related concerns for humanity, should be examined carefully as they are a global voice of young people who want to be involved and be heard in decisions that will impact their lives in the future. After all, young people represent the future of the space industry.

[161] "About the Space Generation Advisory Council." Space Generation Advisory Council. 15 June 2009. http://www.spacegeneration.org/About.
[162] "Declaration of the Space Generation." 27 July 1999. Technical Report of the Space Generation Forum. UN document A/CONF.184/14. 15 June 2009. http://www.spacegeneration.org/downloads/documents/UN/sgf-technical-report-july-1999.pdf.

Annex

The "10 Steps to Achieve Fair and Responsible Use of Outer Space"

As derived from the IAA/ESPI/SWF Conference on 20/21 November 2008 in Vienna

1. Apply a holistic view to the role of space-based solutions in achieving sustainable development (environment, resources, security, knowledge, mobility, energy) and ensure data sharing.
A number of countries lack the means to take full advantage of space applications in environmental security, disaster management, resource management, education and security. Governments need to provide commercial incentives to develop related space applications. Resources, expertise and experiences have to be shared. Given the change in geopolitics in the post-Cold War environment and the resulting increase in commercial actors in space, there is a greater need for cooperation to lower the pressure on the existing resources and the environment.

2. Enhance the use of space in organisations dealing with development and sustainability issues, in particular in the UN system.
The UN has to assume the leadership role that is needed and broaden its use of space applications. In this context it may develop a United Nations' Space Policy. Particularly, the inter-agency dialogue has to be strengthened. Coordination has also to be sought with institutions and organisations outside the UN system, wherever this may lead to increased efficiency in the use of space applications for sustainable development.

3. Strengthen the implementation and application of a harmonised legal regime and create non-binding regulations and standardisations for achieving fairness, responsibility and peace in outer space.
The existing legal regime has to be strengthened through enhanced ratification by States of the treaties. In the course of establishing national space legislation, the States should at an early stage aim at harmonising their regulations in order to avoid "flags of convenience" but maintain a level playing field. Legal non-binding regulations have to involve all relevant actors in order to guarantee fairness and universality.

4. Establish and apply consultation mechanisms, decision-making procedures and enforcement mechanisms on the international level, which guarantee global involvement, including public and private operators and users of outer space.
A fair international system requires making opportunities for all actors to get involved in shaping the future. Consultation and decision-making must not be the privilege of a few actors. Non-governmental space actors (especially the large satellite operators) have to be integrated in a future governance scheme. The framework for envisaging such new ways could be provided by a future UNISPACE conference.

5. Develop fair traffic management rules for all actors, the present ones as well as newcomers. Aim at a common understanding between civilian and military users under mutual goals.
Governments together with private operators have to consult on establishing a fair traffic management system, providing timely information on possible interference as well as verification of mitigation plans and service availability at all times. A forum for discussion for civilian and military actors has to be established, which breaks down barriers between these two groups in data sharing issues and helps to distinguish common and mutual grounds for cooperation.

6. Protect the space environment and make clear that the newcomers must avoid the mistakes of the earlier users. Also protect the planetary environment.
A mechanism has to be established, acquainting newcomers with the experience and necessary data of the existing space-faring countries to maintain a safe space environment. In this regard countries especially have to share their experience in implementing the space debris mitigation guidelines. The objective of such an effort shall be to establish customary procedures for protecting near-Earth space and the planetary environment.

7. Guarantee the fair and equitable use of the frequency spectrum for all space activities to all actors.
The limited natural resource of the frequency spectrum (together with the orbital positions) has to be used in a fair and equitable manner. Mechanisms established by the ITU have long been models for achieving fairness in international relations. This status has to be maintained and further developed amidst all the geopolitical changes.

8. Avoid an arms race in outer space and prevent space from becoming an area of (armed) conflict.
Space is increasingly used by militaries and more and more States depend on their space-assets in war fighting. Given this high vulnerability resulting from increased

military dependence on space, avoiding an arms race in Space is of the greatest importance. Space cannot become the next high-ground.

9. Create mechanisms that allow emerging space powers and non-space-faring countries also to participate in the human exploration of space.
Not all countries can engage in space exploration by contributing technology or money. Yet space exploration should to be an endeavour of the whole humankind. Mechanisms be established to allow the "have-nots" in this field to benefit from the motivation and the spirit that space exploration can provide.

10. Look at space in a long-term perspective and involve the youth.
The perceptions, ideas and wishes of the youth should be carefully evaluated and taken into account when formulating new policies. Developing a long-term perspective that includes the next generation is of utmost importance. The young generation interested in space today have to become the leaders in technology for tomorrow in order to increase the benefits that space applications and exploration bring to humankind.

Draft code of conduct for outer space activities

As published by the Council of the European Union on 3 December 2008

Preamble

The Subscribing States,

Noting that all States should actively contribute to the promotion and strengthening of international cooperation relating to the activities in the exploration and use of outer space for peaceful purposes (hereinafter referred to as outer space activities);

Recognising the need for the widest possible adherence to relevant existing international instruments that promote the peaceful uses of outer space in order to meet emerging new challenges;

Convinced that the use of existing space technology, space telecommunications, and their applications, has important consequences in the economic, social and cultural development of nations;

Further recognising that space capabilities – including associated ground and space segments and supporting links – are vital to national security and to the maintenance of international peace and security;

Recalling the initiatives aiming at promoting a peaceful, safe and secure outer space environment, through international cooperation;

Recalling the importance of developing transparency and confidence-building measures for activities in outer space;

Taking into account that space debris could constitute a threat to outer space activities and potentially limit the effective deployment and exploitation of associated space capabilities;

Reaffirming their commitment to resolve any conflict concerning actions in space by peaceful means;

Recognising that a comprehensive approach to safety and security in outer space should be guided by the following principles: (i) freedom of access to space for all for peaceful purposes, (ii) preservation of the security and integrity of space objects in orbit and (iii) due consideration for the legitimate defence interests of States;

Conscious that a comprehensive code, including transparency and confidence-building measures could contribute to promoting common and precise understandings;

Adopt the following Code (hereinafter referred to as "the Code").

I. Core principles and objectives

1. Purpose and scope

1.1. The purpose of the present code is to enhance the safety, security and predictability of outer space activities for all.
1.2. The present Code is applicable to all outer space activities conducted by a Subscribing State or jointly with other State(s) or by non-governmental entities under the jurisdiction of a Subscribing State, including those activities within the framework of international intergovernmental organisations.
1.3. This Code, in codifying new best practices, contributes to transparency and confidence-building measures and is complementary to the existing framework regulating outer space activities.
1.4. Adherence to this Code and to the measures contained in it is voluntary and open to all States.

2. General principles

The Subscribing States resolve to abide by the following principles:

- the freedom of access to, exploration and use of outer space and exploitation of space objects for peaceful purposes without interference, fully respecting the security, safety and integrity of space objects in orbit;
- the inherent right of individual or collective self-defence in accordance with the United Nations Charter;
- the responsibility of States to take all the appropriate measures and cooperate in good faith to prevent harmful interference in outer space activities;
- the responsibility of States, in the conduct of scientific, commercial and military activities, to promote the peaceful exploration and use of outer space and take all the adequate measures to prevent outer space from becoming an area of conflict;

3. Compliance with and promotion of treaties, conventions and other commitments relating to outer space activities

3.1. The Subscribing States reaffirm their commitment to:

- the existing legal framework relating to outer space activities;
- making progress towards adherence to, and implementation of:

(a) the existing framework regulating outer space activities, inter alia:

- the Treaty on Principles Governing the Activities of States in the Exploration and Use of Outer Space, including the Moon and Other Celestial Bodies (1967);
- the Agreement on the Rescue of Astronauts, the Return of Astronauts and the Return of Objects Launched into Outer Space (1968);
- the Convention on International Liability for Damage Caused by Space Objects (1972);
- the Convention on Registration of Objects Launched into Outer Space (1975);
- the Constitution and Convention of the International Telecommunications Union and its Radio Regulations (2002);
- the Treaty banning Nuclear Weapon Tests in the Atmosphere, in Outer Space and under Water (1963) and the Comprehensive Nuclear Test Ban Treaty (1996);
- the International Code of Conduct against Ballistic Missile Proliferation (2002).

(b) declarations and Principles, inter alia:

- the Declaration of Legal Principles Governing the Activities of States in the Exploration and Use of Outer Space as stated in UNGA Resolution 1962 (XVIII);
- the Principles Relevant to the Use of Nuclear Power Sources in Outer Space as stated in UNGA Resolution 47/68;
- the Declaration on International Cooperation in the Exploration and Use of Outer Space for the Benefit and in the Interest of All States, Taking into Particular Account the Needs of Developing Countries as stated in UNGA Resolution 51/122;
- the Recommendations on the Practice of States and International Organisations in Registering Space Objects as stated in UNGA Resolution 62/101;
- the Space Debris Mitigation Guidelines of the United Nations Committee for the Peaceful Uses of Outer Space as stated in UNGA Resolution 62/217.

3.2. The Subscribing States also reiterate their support to encourage coordinated efforts in order to promote universal adherence to the above mentioned instruments.

II. General measures

4. Measures on space operations

4.1. The Subscribing States will establish and implement national policies and procedures to minimise the possibility of accidents in space, collisions between space objects or any form of harmful interference with other States' right to the peaceful exploration and use of outer space.

4.2. The Subscribing States will, in conducting outer space activities:

- refrain from any intentional action which will or might bring about, directly or indirectly, the damage or destruction of outer space objects unless such action is conducted to reduce the creation of outer space debris and/or justified by imperative safety considerations;
- take appropriate steps to minimise the risk of collision;
- abide by and implement all International Telecommunications Union recommendations and regulations on allocation of radio spectra and orbital assignments.

4.3. When executing manoeuvres of space objects in outer space, for example to supply space stations, repair space objects, mitigate debris, or reposition space objects, the Subscribing States agree to take all reasonable measures to minimise the risks of collision.

4.4. The Subscribing States resolve to promote the development of guidelines for space operations within the appropriate fora for the purpose of protecting the safety of space operations and long term sustainability of outer space activities.

5. Measures on space debris control and mitigation

In order to limit the creation of space debris and reduce its impact in outer space, the Subscribing States will:

- refrain from intentional destruction of any on-orbit space object or other harmful activities which may generate long-lived space debris;
- adopt, in accordance with their national legislative processes, the appropriate policies and procedures in order to implement the Space Debris Mitigation Guidelines of the United Nations Committee for the Peaceful Uses of Outer Space as endorsed by UNGA Resolution 62/217.

III. Cooperation mechanisms

6. Notification of outer space activities

6.1. The Subscribing States commit to notify, in a timely manner, to the greatest extent feasible and practicable, all potentially affected Subscribing States on the outer space activities conducted which are relevant for the purposes of this Code, inter alia:

- the scheduled manoeuvres which may result in dangerous proximity to space objects;
- orbital changes and re-entries, as well as other relevant orbital parameters;
- collisions or accidents which have taken place;
- the malfunctioning of orbiting space objects with significant risk of re-entry into the atmosphere or of orbital collision.

6.2. The Subscribing States reaffirm their commitment to the Principles Relevant to the Use of Nuclear Power Sources in Outer Space as stated in UNGA Resolution 47/68.

7. Registration of space objects

The Subscribing States undertake to register space objects in accordance with the Convention on Registration of Objects launched in Outer Space and to provide the United Nations Secretary-General with the relevant data as set forth in this Convention and in the Recommendations on the Practice of States and International Organisations in Registering Space Objects as stated in UNGA Resolution 62/101.

8. Information on outer space activities

8.1. The Subscribing States resolve to share, on an annual basis, and, where available, information on:

- national space policies and strategies, including basic objectives for security and defence related activities;
- national space policies and procedures to prevent and minimise the possibility of accidents, collisions or other forms of harmful interference;
- national space policies and procedures to minimise the creation of space debris;
- efforts taken in order to promote universal adherence to legal and political regulatory instruments concerning outer space activities.

8.2. The Subscribing States may also consider providing timely information on space environmental conditions and forecasts to other Subscribing States or private entities through their national space situational awareness capabilities.

9. Consultation mechanism

9.1. Without prejudice to existing consultation mechanisms provided for in Article IX of the Outer Space Treaty of 1967 and in Article 56 of the ITU

Constitution, the Subscribing States have decided on the creation of the following consultation mechanism:

- A Subscribing State with reason to believe that certain outer space activities conducted by one or more Subscribing State(s) are, or may be, contrary to the purposes of the Code may request consultations with a view to achieving acceptable solutions regarding measures to be adopted in order to prevent or minimise the inherent risks.
- The Subscribing States involved in a consultation process will decide on a timeframe consistent with the timescale of the identified risk triggering the consultations.
- Any other Subscribing State which may be affected by the risk and requests to take part in the consultations will be entitled to take part.
- The Subscribing States participating in the consultations shall seek solutions based on an equitable balance of interests.

9.2. In addition, the Subscribing States may propose to create a mechanism to investigate proven incidents affecting space objects. The mechanism, to be agreed upon at a later stage, could be based on national information and/or national means of investigation provided on a voluntary basis by the Subscribing States and on a roster of internationally recognised experts to undertake an investigation.

IV. Organisational aspects

10. Biennial meeting of Subscribing States

10.1. The Subscribing States decide to hold meetings biennially or as otherwise agreed by Subscribing States, to define, review and further develop this Code and ensure its effective implementation. The agenda for such biennial meetings could include: (i) review of the implementation of the Code, (ii) evolution of the Code and (iii) additional measures which appear necessary.

10.2. The decisions will be taken by consensus of the Subscribing States present at the meeting.

11. Central point of contact

A central point of contact shall be nominated among Subscribing States to:

- receive and announce the subscription of additional States;
- maintain the electronic information-sharing system;
- serve as secretariat at the biennial meetings of Subscribing States;
- carry out other tasks as agreed by Subscribing States.

12. Outer space activities database

The Subscribing States will create an electronic database to:

- collect and disseminate notifications and information submitted in accordance with the provisions of this Code;
- channel requests for consultations.

Space Debris mitigation guidelines

As stated in the Annex of Supplement No. 20 to the Official Records of the sixty-second session of the Committee on the Peaceful Uses of Outer Space (A/62/20)

1. Background

Since the Committee on the Peaceful Uses of Outer Space published its Technical Report on Space Debris in 1999, it has been a common understanding that the current space debris environment poses a risk to spacecraft in Earth orbit. For the purpose of this document, space debris is defined as all man-made objects, including fragments and elements thereof, in Earth orbit or re-entering the atmosphere, that are non-functional. As the population of debris continues to grow, the probability of collisions that could lead to potential damage will consequently increase. In addition, there is also the risk of damage on the ground, if debris survives Earth's atmospheric re-entry. The prompt implementation of appropriate debris mitigation measures is therefore considered a prudent and necessary step towards preserving the outer space environment for future generations.

Historically, the primary sources of space debris in Earth orbits have been (a) accidental and intentional break-ups which produce long-lived debris and (b) debris released intentionally during the operation of launch vehicle orbital stages and spacecraft. In the future, fragments generated by collisions are expected to be a significant source of space debris.

Space debris mitigation measures can be divided into two broad categories: those that curtail the generation of potentially harmful space debris in the near term and those that limit their generation over the longer term. The former involves the curtailment of the production of mission-related space debris and the avoidance of break-ups. The latter concerns end-of-life procedures that remove decommissioned spacecraft and launch vehicle orbital stages from regions populated by operational spacecraft.

2. Rationale

The implementation of space debris mitigation measures is recommended since some space debris has the potential to damage spacecraft, leading to loss of mission, or loss of life in the case of manned spacecraft. For manned flight orbits, space debris mitigation measures are highly relevant due to crew safety implications.

A set of mitigation guidelines has been developed by the Inter-Agency Space Debris Coordination Committee (IADC), reflecting the fundamental mitigation elements of a series of existing practices, standards, codes and handbooks developed by a number of national and international organizations. The Committee on the Peaceful Uses of Outer Space acknowledges the benefit of a set of high-level qualitative guidelines, having wider acceptance among the global space community. The Working Group on Space Debris was therefore established (by the Scientific and Technical Subcommittee of the Committee) to develop a set of recommended guidelines based on the technical content and the basic definitions of the IADC space debris mitigation guidelines, and taking into consideration the United Nations treaties and principles on outer space.

3. Application

Member States and international organizations should voluntarily take measures, through national mechanisms or through their own applicable mechanisms, to ensure that these guidelines are implemented, to the greatest extent feasible, through space debris mitigation practices and procedures.

These guidelines are applicable to mission planning and the operation of newly designed spacecraft and orbital stages and, if possible, to existing ones. They are not legally binding under international law.

It is also recognized that exceptions to the implementation of individual guidelines or elements thereof may be justified, for example, by the provisions of the United Nations treaties and principles on outer space.

4. Space debris mitigation guidelines

The following guidelines should be considered for the mission planning, design, manufacture and operational (launch, mission and disposal) phases of spacecraft and launch vehicle orbital stages:

Guideline 1: Limit debris released during normal operations
Space systems should be designed not to release debris during normal operations. If this is not feasible, the effect of any release of debris on the outer space environment should be minimized.

During the early decades of the space age, launch vehicle and spacecraft designers permitted the intentional release of numerous mission-related objects into Earth orbit, including, among other things, sensor covers, separation mechanisms and deployment articles. Dedicated design efforts, prompted by the recognition of the threat posed by such objects, have proved effective in reducing this source of space debris.

Guideline 2: Minimize the potential for break-ups during operational phases
Spacecraft and launch vehicle orbital stages should be designed to avoid failure modes which may lead to accidental break-ups. In cases where a condition leading to such a failure is detected, disposal and passivation measures should be planned and executed to avoid break-ups.

Historically, some break-ups have been caused by space system malfunctions, such as catastrophic failures of propulsion and power systems. By incorporating potential break-up scenarios in failure mode analysis, the probability of these catastrophic events can be reduced.

Guideline 3: Limit the probability of accidental collision in orbit
In developing the design and mission profile of spacecraft and launch vehicle stages, the probability of accidental collision with known objects during the system's launch phase and orbital lifetime should be estimated and limited. If available orbital data indicate a potential collision, adjustment of the launch time or an on-orbit avoidance manoeuvre should be considered.

Some accidental collisions have already been identified. Numerous studies indicate that, as the number and mass of space debris increase, the primary source of new space debris is likely to be from collisions. Collision avoidance procedures have already been adopted by some Member States and international organizations.

Guideline 4: Avoid intentional destruction and other harmful activities
Recognizing that an increased risk of collision could pose a threat to space operations, the intentional destruction of any on-orbit spacecraft and launch vehicle orbital stages or other harmful activities that generate long-lived debris should be avoided.

When intentional break-ups are necessary, they should be conducted at sufficiently low altitudes to limit the orbital lifetime of resulting fragments.

Guideline 5: Minimize potential for post-mission break-ups resulting from stored energy
In order to limit the risk to other spacecraft and launch vehicle orbital stages from accidental break-ups, all on-board sources of stored energy should be depleted or

made safe when they are no longer required for mission operations or post-mission disposal.

By far the largest percentage of the catalogued space debris population originated from the fragmentation of spacecraft and launch vehicle orbital stages. The majority of those break-ups were unintentional, many arising from the abandonment of spacecraft and launch vehicle orbital stages with significant amounts of stored energy. The most effective mitigation measures have been the passivation of spacecraft and launch vehicle orbital stages at the end of their mission. Passivation requires the removal of all forms of stored energy, including residual propellants and compressed fluids and the discharge of electrical storage devices.

Guideline 6: Limit the long-term presence of spacecraft and launch vehicle orbital stages in the low-Earth orbit (LEO) region after the end of their mission
Spacecraft and launch vehicle orbital stages that have terminated their operational phases in orbits that pass through the LEO region should be removed from orbit in a controlled fashion. If this is not possible, they should be disposed of in orbits that avoid their long-term presence in the LEO region.

When making determinations regarding potential solutions for removing objects from LEO, due consideration should be given to ensuring that debris that survives to reach the surface of the Earth does not pose an undue risk to people or property, including through environmental pollution caused by hazardous substances.

Guideline 7: Limit the long-term interference of spacecraft and launch vehicle orbital stages with the geosynchronous Earth orbit (GEO) region after the end of their mission
Spacecraft and launch vehicle orbital stages that have terminated their operational phases in orbits that pass through the GEO region should be left in orbits that avoid their long-term interference with the GEO region.

For space objects in or near the GEO region, the potential for future collisions can be reduced by leaving objects at the end of their mission in an orbit above the GEO region such that they will not interfere with, or return to, the GEO region.

5. Updates

Research by Member States and international organizations in the area of space debris should continue in a spirit of international cooperation to maximize the benefits of space debris mitigation initiatives. This document will be reviewed and may be revised, as warranted, in the light of new findings.

6. Reference

The reference version of the IADC space debris mitigation guidelines at the time of the publication of this document is contained in the annex to document UN Doc. A/AC.105/C.1/L.260.

For more in-depth descriptions and recommendations pertaining to space debris mitigation measures, Member States and international organizations may refer to the latest version of the IADC space debris mitigation guidelines and other supporting documents, which can be found on the IADC website (www.iadc-online.org).

About the authors

William Ailor, Ph.D., is the director of the Center for Orbital and Reentry Debris Studies at The Aerospace Corporation, located in El Segundo, California, USA. He was general chairman of the 2004 and 2007 Planetary Defense Conferences: Protecting Earth from Asteroids and is co-chair of the 1st International Academy of Astronautics Planetary Defense Conference to be held in April 2009. He has worked in the reentry breakup area for over 30 years. He heads the U.S. delegation on ISO's Orbital Debris Coordinating Working Group, which leads the development of international standards to minimize the production of orbital debris, and he contributed to the development of the IAA Position Paper on Space Debris Mitigation and IAA Cosmic Study on Traffic Management Rules for Space Operations. Dr. Ailor has published several articles and professional papers on space traffic control, re-entry breakup, planetary defence and space debris.

Fernand Alby is an engineer and holds a diploma from the Ecole Centrale de Paris. He joined the Flight Dynamics Division of CNES (Centre National d'Etudes Spatiales) in 1982 and participated in positioning and station keeping operations for several geostationary satellites. He then took the responsibility of a department in charge of technical support to the Hermes spaceplane project in the field of atmospheric reentry, in-orbit rendezvous and navigation. He then took the overall responsibility for space debris and space surveillance activities at CNES: this includes research studies, regulatory activities and operations. Fernand ALBY is member of the Steering Group of the IADC (Inter Agency Space Debris Coordination Committee) and is technical support to the French delegation to the UN-COPUOS Scientific and Technical Sub-Commitee. He is author of numerous publications and Member of the International Academy of Astronautics.

Dr. Ciro Arevalo Yepes is the current Chairman of the UN Committee on the Peaceful Uses of Outer Space (COPUOS) as well as the advisor on space matters to the Colombian Ministry of Foreign Affairs. He has worked in the space sector for the past 20 years. He has law degree from Universite de Lausanne, Switzerland, and postgraduate studies in international relations and international public law at the Universidad de los Andes. He is a member of Colombian Bar Association since 1985. Member of the Colombian delegation to various international entities including ITU, IAEA, UNTAT, ILO, CBTO, etc. Dr. Arevalo Yepes was one of the main organizers of the Fourth Space Conference of the Americas held in

Cartagena, Colombia in 2002, during which the Colombian Space Commission has originated. In addition to his diplomatic career, Dr. Arevalo Yepes has dedicated portion of his time to the academics. He was a Dean of the Post Graduate International Business Programme at University of Bogotá, UJTL and a Founder and Director of the Centre of U.S. Studies, CEUS, Bogotá. He authored a number of papers and articles on Space- related matters, Space and Society, and Space and Education.

Ben Baseley-Walker is Legal and Policy Advisor at the Secure World Foundation (SWF). With Secure World Foundation, Mr. Baseley-Walker analyses the international legality of space security issues, facilitates briefings, meetings and engagement among public, private and non-profit members of the space community in Washington, DC, and contributes to the work of various United Nations fora including the UN Committee on the Peaceful Uses of Outer Space, the Conference on Disarmament (CD) and the UN General Assembly. In 2009, Mr. Baseley-Walker delivered the first ever NGO presentation to an informal plenary session of the UN Conference on Disarmament. He is also currently Chair of the Space Generation Advisory Council. Before joining SWF, Mr. Baseley-Walker worked in international relations and international law in East Africa with the Kenya National Commission on Human Rights and the United Nations High Commissioner for Refugees in Nairobi, Kenya. Prior to his work in Kenya, Mr. Baseley-Walker worked as Assistant Project Coordinator for the Ethiopia Legal Project where he liaised between governmental partners in Europe and Ethiopia. Mr. Baseley-Walker studied International and European Law at the Universiteit Van Amsterdam Law School in the Netherlands and Politics at the University of Edinburgh in Scotland. He is a speaker of French, Spanish and basic Mandarin in addition to his native English and a member of the American Society of International Law.

Gérard Brachet was born in October 1944 in Lyon, France. He holds an Engineering degree from the *Ecole Nationale Supérieure d'Aéronautique* (1967) and a Master of Sciences in Aeronautics and Astronautics from the University of Washington (1968). From 1970 to 1982, he held various positions at *Centre National d'Etudes Spatiales* (CNES), *the French Space Agency.* He was also Chairman of the European Space Agency's Programme Board for Earth observation from 1979 to 1981. From 1982 to 1994, he was Chairman and Chief executive officer of SPOT IMAGE, the company set up by CNES to develop the market for remote sensing imagery from the SPOT series of satellites. In 1995, he went back to CNES as Director for programmes, planning and industrial policy. Subsequently, he was Director General of CNES from 1997 to 2002. In addition, he was Chairman of the international "Committee on Earth Observation Satellites (CEOS)" in 1997. From 2003 onwards, Gérard Brachet has been a consultant on space policy issues and space applications, in particular to the European Commission and the European Space Agency. From June 2006 to June 2008, he served as chairman of the United Nations Committee for the Peaceful Uses of Outer Space (UN COPUOS). He is also President of the Air and Space Academy (*Académie de l'Air et de l'Espace*), Vice President of the International Astronautical Federation (IAF) and chairman of its Committee on Liaison with International Organizations and Developing Nations (CLIODN). Moreover, he is a member of the International Academy of Astronautics. Gérard Brachet holds numerous awards and titles, such as *Officier de l'Ordre National du Mérite* and *Officier de la Légion d'Honneur.*

Driss El Hadani is the Director of the Royal Centre for Remote Sensing (CRTS) and chairman of the National Remote Sensing Committee. First Vice-president (2000–2004) of the United Nations Committee for the Peaceful Uses of Outer Space (COPUOS). Vice-president (2002–2004) of the International Astronautic Federation (IAF). Member of the UN-OOSA Working Group on strategy of integration of the space techniques for risk and disasters natural management. Member of the joint UNESCO/African Union high-level scientific group on the role of satellite remote sensing for sustainable development. Member of the Scientific committee of the African Centre for Catastrophe Risks created (in April 2005) by the African Insurance Organisation.

Driss El Hadani is a Hydrogeologist engineer. He holds a DESS Diploma in Space Remote Sensing Applied to Environment, University of Paris VI, France, a MBA of the Higher Institute of the Trade and Business Administration (ISCAE), Casablanca, Morocco, and a Bachelor in Law, University of Rabat, Morocco.

Giovanni Gasparini, Senior Fellow, Defence and Security, Istituto Affari Internazionali, Rome, Italy, and Co-Director, Transatlantic Programme on ESDP, Istituto Affari Internazionali, graduated in Economics: he joined IAI in 1999; his main areas of research are ESDP, NATO, space, aerospace and defence industry, defence economics and transatlantic relations. He has published extensively on these subjects and he is member of the Editorial Board of the "The International Spectator". He served in the Italian Air Force, Military Researcher at CeMiSS, Military Centre for Strategic Studies of the Italian MoD. Since 2001 he is consultant to CeMiSS, where he contributes regularly to the Monthly Observatory on European Defence and Defence Industry issues. He has been Visiting Fellow at the EU-ISS in Paris, at SIPRI in Stockholm, at FRS in Paris, at ISIS-Europe in Brussels, member of the US State Department IVP Programme, as well as manager and participant of numerous European defence, security and space projects for the EC, EDA, EP and ESA.

Niklas Hedman is Chief of Committee Services and Research Section of the United Nations Office for Outer Space Affairs. He is Secretary of the Committee on the Peaceful Uses of Outer Space and its Scientific and Technical Subcommittee and Legal Subcommittee. He is also Secretary of the United Nations Inter-Agency Meeting on Outer Space Activities, which is the central coordination mechanism for space-related activities in the United Nations system. Before joining the United Nations in 2006, he worked in the Swedish Ministry for Foreign Affairs, particularly in the areas of ocean affairs and law of the sea, space law and space policy, as well as disarmament and non-proliferation, including PAROS and the Hague Code of Conduct Against Ballistic Missile Proliferation (HCOC). He represented Sweden to COPUOS for 10 years and held various positions, including Chairman of the UNISPACE III + 5 report A/59/174. Mr. Hedman represented Sweden to the final rounds of negotiations on the International Space Station Intergovernmental Agreement (ISS-IGA), and was chief negotiator to the governmental framework agreement on space cooperation between Sweden and the United States of America. He is a member of the International Space Law Committee of the International Law Association (ILA), International Institute of Space Law (IISL) and International Academy of Astronautics (IAA).

Theresa Hitchens, Director, UN Institute for Disarmament Research, Geneva, and Editor of *Defense News* from 1998 to 2000, has had a long career in journalism, with a focus on military, defence industry and NATO affairs. Her time at *Defense News* included five years as the newspaper's first Brussels bureau chief, from 1989 to 1993. From 1983 to 1988, she worked at Inside Washington Publishers on the group's environmental and defence-related newsletters, covering issues from nuclear waste to electronic warfare to military space. She has had a long interest in security policy and politics, having served internships with Sen. John Glenn, D-Ohio, and with the NATO Parliamentary Assembly in Brussels. Most recently, she was director of research at the British American Security Information Council, a think tank based in Washington and London. Having been director of the Center for Defense Information (CDI), Hitchens lead CDI's Space Security Project in cooperation with the Secure World Foundation. In 2009 she took up her post as Director of the UN Institute for Disarmament Research at Geneva. The author of "Future Security In Space: Charting a Cooperative Course," she also continues to write on space and nuclear arms control issues for a number of outside publications. She serves on the editorial board of The Bulletin of the Atomic Scientists and is a member of

Women in International Security and the International Institute for Strategic Studies.

Alex Karl holds a degree from the University of Applied Sciences Aachen, Germany, in Aerospace Engineering and a Postgraduate Master in Space Exploration and Development Systems from the Politecnico di Torino, Italy. He has worked for the Max-Planck Instiitute for Radioastronomy in Bonn, Germany and Austrian Aerospace in Vienna, Austria, and is currently working for Space Applications Services as an Operations Engineer at the Belgian User Support and Operations Center in Brussels, Belgium. As a volunteer, he has been the Co-Chair of the Space Generation Advisory Council and is now the Project Coordinator of the organisation.

Agnieszka Lukaszczyk is, since September 2006, the Executive Officer of the Space Generation Advisory Council (SGAC). Starting May 2008 she has also been working as a space policy consultant at the Secure World Foundation (SWF). In addition, during the period of September 2006 to June 2008, Agnieszka worked at the European Space Policy Institute (ESPI). Agnieszka also serves as the Vice President – Operations for the World Space Week. She holds a Master's degree from the American University's School of International Service in International Politics and a Bachelor's degree in Political Science from the University of Tennessee. She also studied at the Universite Catholique de Louvain in Brussels, Belgium, the Jagiellonian University in Krakow, Poland, and the World Trade Institute in Berne, Switzerland. She gained professional experience at the Political Section of the Polish Embassy in Washington, DC, American Electronics Association in Brussels, European Department of the Polish Senate in Warsaw and the Warsaw Business Journal.

Dr. Peter Martinez is the Division Head for Space Science and Technology in the South African Astronomical Observatory, a national research facility under the South African National Research Foundation. He has been a member of the South African delegation to UN COPUOS since 2002. Dr. Martinez also has extensive knowledge of space affairs in Africa. In 2007 he was the principal organiser of the 2nd African Leadership Conference on Space Science and Technology for Sustainable Development, held in Pretoria from 3 to 5 October 2007. Dr. Martinez is President of the Astronomical Society of Southern Africa and President of the International Astronomical Union's Commission 25 on Photometry and Polarimetry. He serves on the Council of the South African Institute of Physics. He is a Fellow of the Royal Astronomical Society (UK) and a COSPAR Associate. He has authored or co-authored over 150 specialist research publications and is an Associate Editor of the international journal *Advances in Space Research*.

Valérie Vicky Miranda is a Junior Researcher in the Security and Defence area at the International Affairs Institute (IAI, Rome). She holds a Bachelor's Degree in Political Sciences and a Master's Degree in International Relations from the University of Roma Tre, where she is currently collaborating with the Chair of European Policies for Security and Development, Faculty of Political Sciences. She has worked at the Italian Ministry of Foreign Affairs (Directorate-General for Development Cooperation) and lectured in several workshops dealing with international cooperation. Her research interests include the EU CFSP and the ESDP, the EU space policy, the relationship between security and defence and R&T, dual-use technologies.

Dr. Luboš Perek is emeritus of the Astronomical Institute of the Academy of Sciences of the Czech Republic. Before, he served as Head of Department and Director of the Institute. In his previous career he was Chief of the UN Outer Space Affairs Division. He is Member of the Czech Learned Society, of the International Astronomical Union, of the European Astronomical Society, Honorary Member of the Czech Astronomical Society, Member of the International Academy of Astronautics, of the Deutsche Akademie der Naturforscher Leopoldina, of the Academie Nationale de l'Air et de l'Espace and of the International Institute of Space Law. He served as President of the International Astronautical Federation and as member of the Board of Directors of the IISL. He is member of the Editorial Board of Space Policy. He was member of the delegation of the Czech Republic to the UN Committee on the Peaceful Uses of Outer Space and its Scientific and Technical Subcommittee. He received the Kepler Medal, the Nagy Ernő Medal, the Nušl Award, Medals of the City of Paris and of Collège de France, the Prix Janssen of the Société Astronomique de France. He is Dr. h.c. of the Masaryk University. He is the author of 40 papers on astronomy, the co-author of the Catalogue of Galactic Planetary Nebulae and author of about 100 papers on space debris, on management of outer space and on the geostationary orbit.

Dr. Wolfgang Rathgeber from Germany studied Electrical Engineering at the universities of Erlangen (Germany), Campinas (Brazil) and Valencia (Spain). He has been with the German Aerospace Centre (DLR) since 1999. Until 2003, he was a doctorate student at the Institute of Radio Frequency Engineering and Radar Systems in Oberpfaffenhofen (Germany) with a thesis on high resolution signal processing for Synthetic Aperture Radar (SAR). From 2003 until 2005, he was scientific staff member at the European Academy of Technology Assessment Bad

Neuenahr-Ahrweiler (Germany). Since 2005 he has been seconded to the European Space Policy Institute (ESPI) in Vienna (Austria), where he is in charge of security topics.

Prof. Dr. Kai-Uwe Schrogl is the Director of the European Space Policy Institute (ESPI) in Vienna, Austria, since 2007. Before, he was Head Corporate Development and External Relations Department in the German Aerospace Center (DLR). In his previous career he worked with the German Ministry for Post and Telecommunications and the German Space Agency (DARA). He has been delegate to numerous international forums and recently served as the chairman of various European and global committees (ESA International Relations Committee, UNCOPUOS working groups). Kai-Uwe Schrogl has published nine books and more than 100 articles, reports and papers in the fields of space policy and law as well as telecommunications policy. He is editor of the "Yearbook on Space Policy" and the book series "Studies in Space Policy" both published at SpringerWienNewYork as well as member in editorial boards of international journals in the field of space policy and law (Acta Astronautica, Space Policy, Zeitschrift für Luft- und Weltraumrecht, Studies in Space Law/Nijhoff). Kai-Uwe Schrogl is Member of the Board of Directors of the International Institute of Space Law, Member of the International Academy of Astronautics (chairing its Commission on policy, economics and law) and the Russian Academy for Cosmonautics. He holds a doctorate degree in political science, lectures international relations at Tübingen University, Germany (as a Honorarprofessor) and has been a regular guest lecturer i.a. at the International Space University and the Summer Courses of the European Centre for Space Law.

Prof. Dr. Kazuto Suzuki is Associate Professor of International Political Economy at Public Policy School of Hokkaido University, Japan. He graduated from Department of International Relations, Ritsumeikan University, and received Ph.D. from Sussex European Institute, University of Sussex, England. He has worked in the University of Tsukuba from 2000 to 2008 and recently moved to Hokkaido University. His research subject is focused on European integration and transformation of national policies with a perspective on security, technology and economy. He has been working on several policy issues such as European space collaboration, defence industry and export control on dual-use technology, but he is also interested in theoretical study on European integration. As an expert of space policy, he has been working as an advisor for Space Development Committee of Liberal Democratic Party of Japan, Society of Japanese Aerospace Industry and policy researcher for JAXA. He has published number of articles and books, both in Japanese and English, including *Policy Logics and Institutions of European Space Collaboration* (Ashgate, 2003), "Administrative Reforms and Policy Logics of Japanese Space Policy", *Space Policy* (Vol. 22, no. 1, 2005), "Transforming Japan's Space Policy-making", *Space Policy* (Vol. 23, no. 2, 2007), "Space Policy of the European Union", in Shoji (ed.) *The Path and Vector of European Integration* (Keio Univ. Press, 2007, Japanese), *Space: Japan's New Security Agenda* (Research Institute of Peace and Security, 2007), "Basic Law for Space Activities: A New Space Policy for Japan for the 21st Century" in Schrogl *et al.* (eds.) *Yearbook on Space Policy 2006/2007* (SpringerWienNewYork, 2008), "Arms Export Policy in Europe", *Journal of World Affairs* (Vol. 56, No. 3, 2008, Japanese), "Japanese Steps toward Regional and Global Confidence Building" in Logsdon and Moltz (eds.) *Collective Security in Space: Asian Perspectives* (Space Policy Institute, The George Washington University, 2008), "Space and Security: Japanese Perspective", *Penser les Ailes françaises* (no. 16, 2008).

Richard Tremayne-Smith, Former Head of Space Environment & International Relations at the BNSXC. Apprentice at the Royal Aircraft Establishment, Farnborough, from the mid-1960s. Assignments included work in Avionics Department (MRCA) & Space Department on the UK X3 or Prospero satellite that was later launched on Black Arrow. During the 1970s and 1980s work on scientific computer systems and research leading to European projects on advanced architectures, the scope of work included parallel, AI and knowledge-based systems. Through the 1990s to date, work on a wide range of space-related areas that included space transportation and propulsion, manned spaceflight, small satellites (MOSAIC Programme), international relations and UN issues, as well as the space environment with emphasis on Space Debris, near-Earth objects (NEOs) and Space Surveillance issues. Providing technical support to the UK licensing process; initiating the review of the implementation of the UK Outer Space Act and taking it forward with due account for the provisions of the outer space treaties and principles.
Qualifications: B.Sc., Ceng., MIET.

Dr. Ray A. Williamson is Executive Director of Secure World Foundation, a private operating foundation with headquarters in Superior, Colorado. He was formerly Research Professor of Space Policy and International Affairs in the Space Policy Institute, The George Washington University. At the institute, he had led several studies of security issues in space and on the socioeconomic benefits of Earth science and space weather research. Ray is also an external faculty member of the International Space University (ISU), Illkirch, France, teaching general space policy and remote sensing for the ISU Masters and Space Studies programmes. He is editor of *Imaging Notes* and serves on the editorial board of the journal Space Policy. As a member of the International Academy of Astronautics, Dr. Williamson serves on Commission Five: Space Policies, Law & Economics. He is the author of more than 100 articles on space policy, remote sensing and space security and author or editor of nine books on outer space, the technologies of historic preservation and American Indian astronomy, myth and ritual. From 1979 to 1995, he was a Senior Analyst and then Senior Associate in the Office of Technology Assessment (OTA) of the U.S. Congress. While at OTA, Dr. Williamson led more than a dozen space policy studies requested by Congressional committees. Ray received his B.A. in physics from the Johns Hopkins University and his Ph.D. in astronomy from the University of Maryland. After 2 years on the astronomy faculty of the University of Hawaii, he taught philosophy, literature, mathematics, physics and astronomy at St. John's College, Annapolis. For the last 5 years he also served as Assistant Dean of the College.

The contributors to the book at the Conference of "Fair and Responsible Use of Space" November 2008 at ESPI; from left: Kai-Uwe Schrogl, Richard Tremayne-Smith, Wolfgang Rathgeber, Gerard Brachet, Giovanni Gasparini, Ciro Arevalo, Ben Baseley-Walker, Ray Williamson, Agnieszka Lucaszczyk, Peter Martinez, Theresa Hitchens, Niklas Hedman, Jean-Michel Contant, William Ailor, Fernand Alby, Lubos Perek, Xavier Pasco, Kazuto Suzuki, Driss El Hadani.

List of acronyms

A

AIAA: American Institute of Aeronautics and Astronautics
APRSAF: Asia-Pacific Regional Space Agency Forum
APSCO: Asia-Pacific Space Cooperation Organisation
ASAT: Anti-satellite
ASI: Agenzia Spaziale Italiana (Italian Space Agency)

B

BASIC: Broad Area Space-Based Imagery Collection
BNSC: British National Space Centre

C

CBERS: China–Brazil Earth Resources Satellites
CD: Conference on Disarmament
CDI: Center for Defense Information
CEOS: Committee on Earth Observation Satellites
CFE: Commercial and Foreign Entities
CNES: Centre National d'Etudes Spatiales (French Space Agency)
CNSA: China National Space Administration
CoC: Code of Conduct
COPUOS: Committee on the Peaceful Uses of Outer Space
COSPAR: Committee on Space Research
CRTS: Centre Royal de Télédétection Spatiale, Morocco
CS: Counterspace

D

DART: Demonstration of Autonomous Rendezvous Technology
DCS: Defensive Counterspace
DLR: Deutsches Zentrum für Luft- und Raumfahrt (German Aerospace Center)
DoD: Department of Defense
DVD: Digital Versatile Disc

E

EADS: European Aeronautic Defence and Space Company
ECOSOC: Economic and Social Council

EDA: European Defence Agency
EMP: Electromagnetic Pulse
ESA: European Space Agency
ESPI: European Space Policy Institute
EU: European Union

F

FAA: Federal Aviation Administration
FRS: Fondation pour la Recherche Stratégique
FY: Feng Yung

G

GEO: Geostationary Orbit
GEOSS: Global Earth Observation System of Systems
GMES: Global Monitoring for Environment and Security
GNSS: Global Navigation Satellite System
GPS: Global Positioning System
GSO: Geosynchronous Orbit

H

HCOC: Hague Code of Conduct on Ballistic Missile Proliferation
HIV: Human Immunodeficiency Virus

I

I3S: Partnership on the Information Society, Science and Spaces Technologies
IAA: International Academy of Astronautics
IAASS: International Association for the Advancement of Space Safety
IADC: Inter-Agency Space Debris Coordination Committee
IAEA: International Atomic Energy Agency
IAF: International Astronautical Federation
ICAO: International Civil Aviation Organization
ICBL: International Campaign to Ban Landmines
ICG: International Committee on Global Navigation Satellite Systems
INSPIRE: Infrastructure for Spatial Information in Europe
IPCC: Intergovernmental Panel on Climate Change
ISO: International Organization for Standardization
ISON: International Space Observation Network
ISRO: Indian Space Research Organisation
ISS: International Space Station

ISU: International Space University
ITAR: International Traffic in Arms Regulations
ITU: International Telecommunication Union

J

JAXA: Japan Aerospace Exploration Agency
JSpOC: Joint Space Operations Centre

L

LDEF: Long Duration Exposure Facility
LEO: Low-Earth Orbit

M

MIB: Mishap Investigation Board
MUSIS: Multinational Satellite-based Imagery System

N

NASA: National Aeronautics and Space Administration
NEO: Near-Earth Objects
NGO: Non-Governmental Organisation
NPO: Non-Profit Organisation
NPoC: National Point of Contact
NPT: Non-Proliferation Treaty
NSAU: National Space Agency of Ukraine

O

OCS: Offensive Counterspace
OECD: Organisation for Economic Co-operation and Development
OOSA: Office for Outer Space Affairs
OST: Outer Space Treaty

P

PAROS: Prevention of an Arms Race in Outer Space
PLA: People's Liberation Army
PPP: Public–Private Partnership

R

RC: Regional Coordinator
RORSAT: Radar Ocean Reconnaissance Satellite

S

SA: Situation Awareness
SAR: Synthetic Aperture Radar
SFA SGAC: Space Force Application Space Generation Advisory Council
SMP: Strategic Master Plan
SOHO: Solar and Heliospheric Observatory
SPOT: Satellite pour l'Observation de la Terre (Earth Observation Satellite)
SSA: Space Situational Awareness
SSN: Space Surveillance Network
SSO: Sun Synchronous Orbit
STM: Space Traffic Management
STS: Space Transportation System
STSC: Scientific and Technical Subcommittee
SUIRG: Satellite Users Interference Reduction Group
SWF: Secure World Foundation

T

TB: Tuberculosis
TM/TC: Telemetry and Telecommand
TV: Television

U

UK: United Kingdom
UN: United Nations
UNCOPUOS: United Nations Committee on the Peaceful Uses of Outer Space
UNESCO: United Nations Educational, Scientific and Cultural Organization
UNISPACE: United Nations Conference on the Exploration and Peaceful Uses of Outer Space
UNOOSA: United Nations Office for Outer Space Affairs
UN-SPIDER: UN Platform for Space-based Information for Disaster Management and Emergency Response
U.S.: United States
USA: United States of America
USAF: United States Air Force
USSR: Union of Soviet Socialist Republics
USSTRATCOM: United States Strategic Command

W

WSO: World Space Organisation
WSSD: World Summit on Sustainable Development
WWII: World War II

List of figures and tables

Figures

1 The general concepts of fairness and responsibility

4 Current questions regarding fair and responsible use of space – summarising introduction

5 Fair rules on orbit – summarising introduction

6 Achieving global engagement – summarising introduction

Tables

4 Current questions regarding fair and responsible use of space – summarising introduction

MIX
Papier aus verantwortungsvollen Quellen
Paper from responsible sources
FSC® C105338

If you have any concerns about our products,
you can contact us on
ProductSafety@springernature.com

In case Publisher is established outside the EU,
the EU authorized representative is:
Springer Nature Customer Service Center GmbH
Europaplatz 3, 69115 Heidelberg, Germany

Printed by Libri Plureos GmbH
in Hamburg, Germany